501

Things <u>YOU</u>
Should Have
Learned
About...

GEOGRAPHY

METRO BOOKS
NEW YORK

An Imprint of Sterling Publishing
387 Part Avenue South
New York, NY 10016

AUTHOR: Sarah Stanbury
SERIES ART DIRECTOR: Clare Barber
SERIES EDITOR: Helena Caldon
DESIGN & EDITING: Quadrum Solutions
PUBLISHER: James Tavendale

IMAGES courtesy of www.shutterstock.com;
www.istockphoto.com and www.clipart.com

ISBN 978-1-4351-4615-0

For information about custom editions, special
sales, and premium and corporate purchases,
please contact Sterling Special Sales at 800-805-5489
or specialsales@sterlingpublishing.com

Printed in China

10 9 8 7 6 5 4 3 2 1

www.sterlingpublishing.com

501

Things <u>YOU</u> Should Have Learned About...

GEOGRAPHY

METRO BOOKS
NEW YORK

CONTENTS

INTRODUCTION

If "explorer" is your middle name and discovering the Earth's various hidden mysteries is your goal, this is the book for you. Explore through the pages of this book and set off on a geographical journey.

Each of the 501 facts included in this book talk about the fascinating aspects of the Earth and its geographical elements—from basic landforms and atmposhere to natural occurences and disasters.

Begin your adventure by learning about maps and then set off on a geographic journey into the ecosystems, water bodies, and landforms, discovering more than you thought you ever could about weather, climate, global issues and their impact on the surroundings, and a lot more.

Don't fret though, this is nothing like those long-winding geography textbooks you had in class. You don't have to worry about memorizing all the amazing facts here, just browse through them like you would a novel. This information will amaze you and compel you to trek from one intriguing fact to another.

What are you waiting for? These geographically rich pages aren't going to flip themselves! Go ahead and start reading...your geographical journey is just beginning...

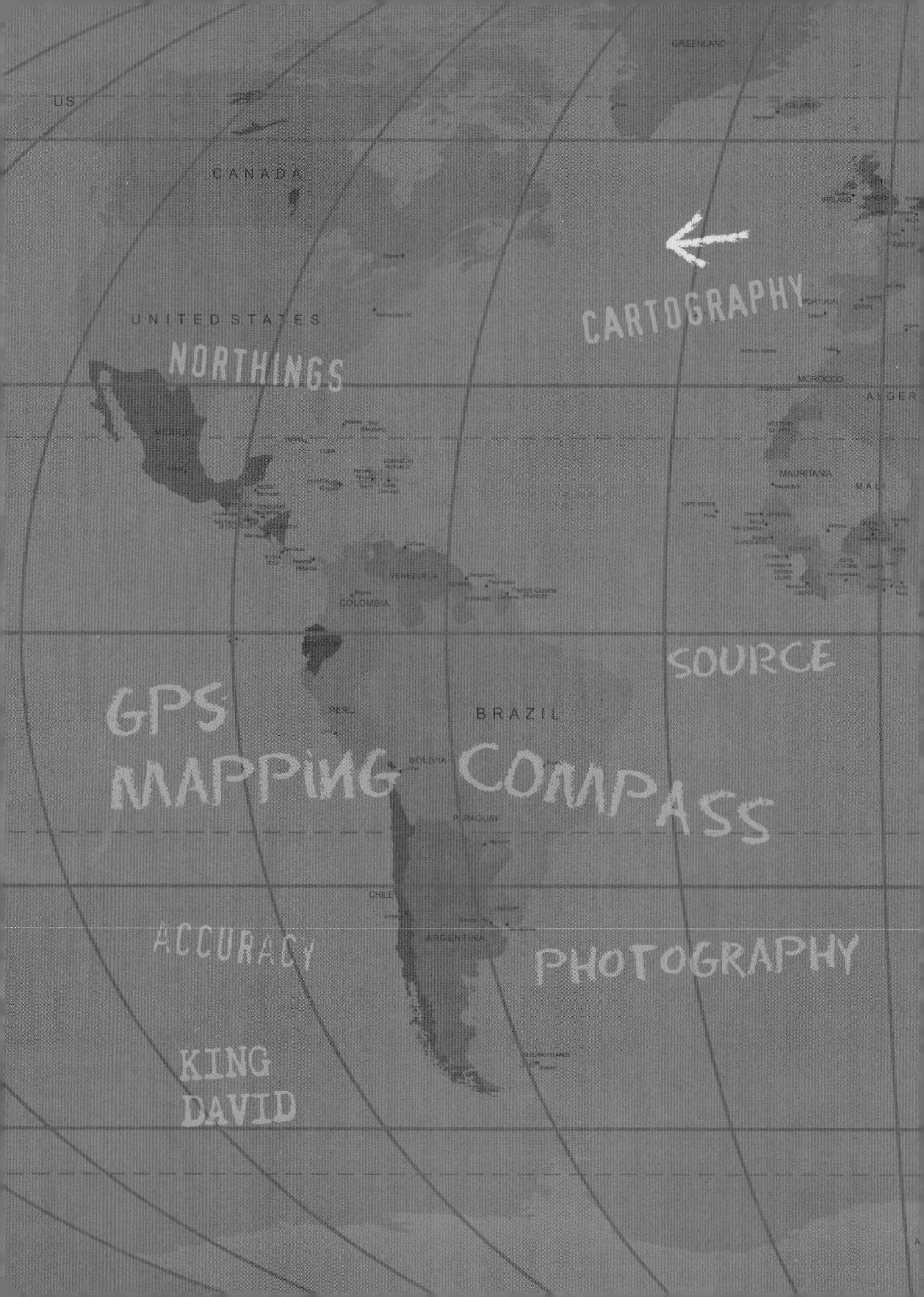

CARTOGRAPHY
NORTHINGS
SOURCE
GPS
MAPPING
COMPASS
ACCURACY
PHOTOGRAPHY
KING
DAVID
GREENLAND
US
CANADA
ICELAND
IRELAND
UNITED STATES
PORTUGAL
SPAIN
MOROCCO
MEXICO
CUBA
MAURITANIA
MALI
VENEZUELA
COLOMBIA
PERU
BRAZIL
BOLIVIA
PARAGUAY
CHILE
ARGENTINA

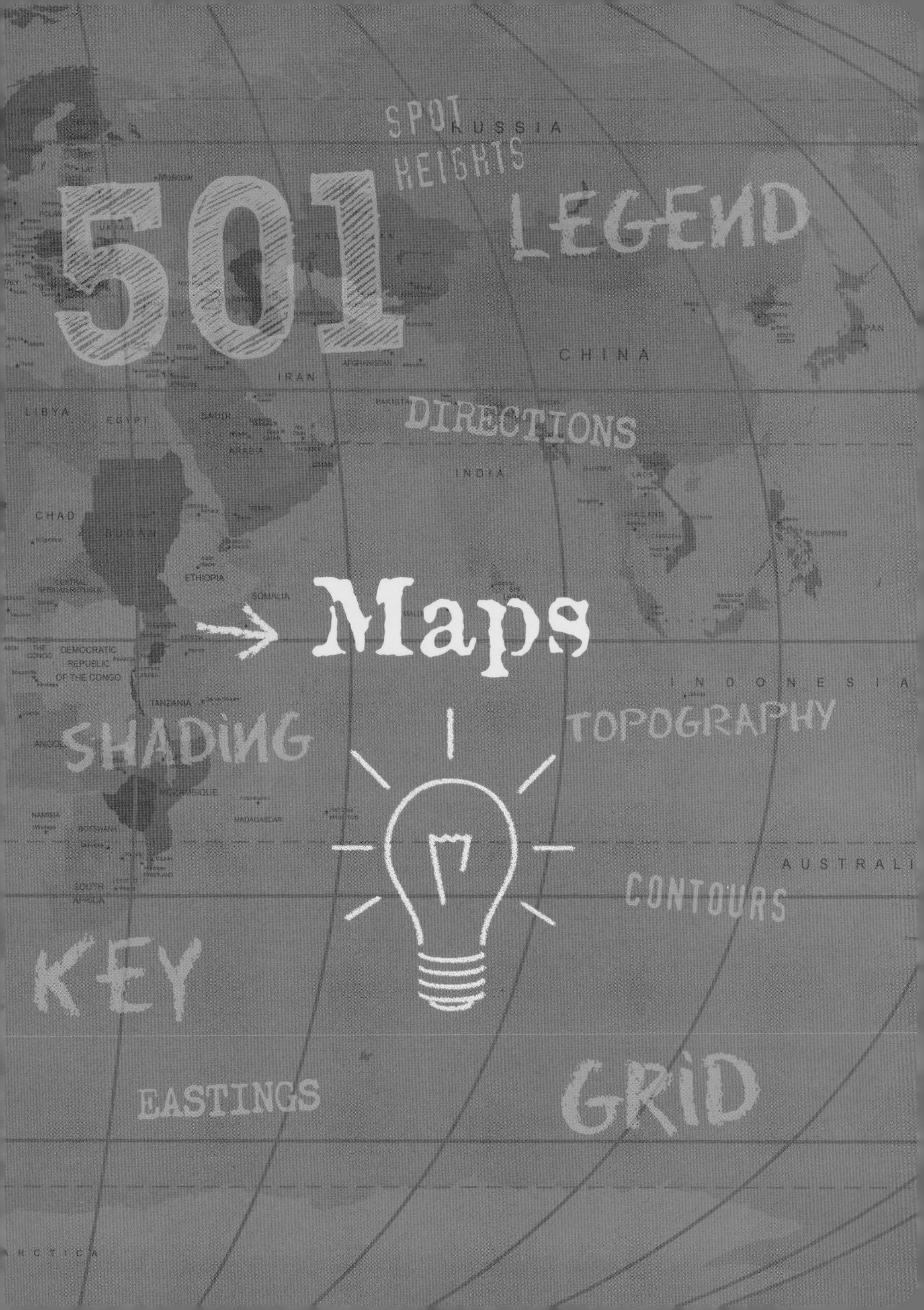
501
Maps
SPOT HEIGHTS
LEGEND
DIRECTIONS
TOPOGRAPHY
SHADING
CONTOURS
KEY
GRID
EASTINGS
RUSSIA
CHINA
JAPAN
IRAN
AFGHANISTAN
LIBYA
EGYPT
INDIA
BURMA
LAOS
CHAD
SUDAN
ETHIOPIA
SOMALIA
CENTRAL AFRICAN REPUBLIC
DEMOCRATIC REPUBLIC OF THE CONGO
TANZANIA
MOZAMBIQUE
NAMIBIA
BOTSWANA
MADAGASCAR
SOUTH AFRICA
INDONESIA
AUSTRALI
PHILIPPINES
ARCTICA

1 GETTING THE BASIC MESSAGE OF MAPS

🎓 **A MAP IS THE MOST BASIC OF GEOGRAPHIC TOOLS** and if there's one thing you should have learned about geography it's how to read a map. Not only will this skill prevent you from getting lost but it'll also help you find your way home when you are lost!

Map makers usually incorporate into their maps a standard set of elements to help you grasp the message. These include:

Title – conveys the subject of the map.

Legend (or key) – contains and defines any symbols on the map.

Scale – provides information about the actual size of the area (typically equating distance in miles and/or kilometers and measurement in inches and/or centimeters) shown on the map.

Orientation – the alignment of the map with respect to cardinal directions. North is usually towards the top of the map ... but not always!

Grid Lines – many maps contain labeled grid lines showing latitude and longitude in order to convey the global context of the mapped area.

Source – most maps provide the source of the information conveyed on the map.

2 FAST FACT...

📖 **A PERSON** makes maps is called a Cartographer.

Gerardus Mercator was a cartographer, mathematician, and philosopher, famously known for creating the 1569 Mercator Projection World Map

3 MAP READING — COMPASS DIRECTIONS

🎓 **THERE ARE SEVERAL THINGS** you need to understand including compass directions, grid references, and the map's key and scale in order to read a map. You need to be able to find features when given a map reference and also be able to describe a feature's location on a map by giving a map reference.

Compass Directions – there are many ways to remember which way is North, South, East and West. Starting at the top and moving clockwise, the directions on a compass or map are:

1. North
2. East
3. South
4. West

You might have learnt a rhyme or a phrase at school to help you remember – if not, here are a couple of options.

'Naughty (North) Elephants (East) Squirt (South) Water (West)' or

'Never (North) Eat (East) Shredded (South) Wheat (West)'.

4 FAST FACT...

📖 **USING A COMPASS IN** interaction with a map is a different skill but definitely one worth exploring to help you navigate safely and accurately in unknown terrain.

5 MAP READING — GRID REFERENCES

A GRID REFERENCE is fundamental when it comes to pinpointing a location on a map. Ordnance Survey maps are divided into numbered squares, which can be used to give a place a four or six-figure grid reference. It's important that you know both four and six-figure grid references.

6 FAST FACT

ALONG the corridor and up the stairs! This will help you remember which way round to read numbers on a grid reference.

Eastings – are lines that run up and down the map. They increase in number the further you move East (or right). You can use them to measure how far to travel East.

Northings – are lines that run across the map horizontally. They increase in number the further you move North (or up the map). You can use them to measure how far to travel north.

Remember:
- numbers along the bottom of the map come first and the numbers up the side of the map come second.

- the four-figure reference 2083 refers to the square to the East of Easting line 20 and North of Northing line 83.

- the six-figure reference 207834 will give you the exact point in the square 2083 - 7/10s of the way across and 4/10s of the way up.

7 FAST FACT...

 WHEN GIVING DIRECTIONS you can provide even more accuracy to your grid reference by stating a nearby landmark or feature.

8 MAP READING – SCALE

SCALE IS THE RELATIONSHIP BETWEEN a distance as measured on a map and the corresponding actual distance on Earth's surface. Calculating the distance between locations and comparing the size of areas are two of the more important functions of maps.

Every map has a single scale although there may be three different ways – bar graph, verbal scale and representative fraction – to tell you what it is.

Bar Graph – looks like a miniature ruler and provides a clear visual reference to the size of the area portrayed on the map.

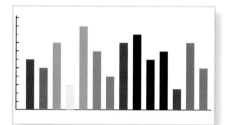

Bar Graph

Verbal Scale (also called statement of scale) – communicates the relationship between map distance and real-world distance in a sentence or sentence-like format. E.g. 'one inch equals one mile'.

Representative Graph

Representative Fraction (RF) – the area shown on a map is a fraction of its actual size. Scale therefore may be indicated, approximately, as a representative fraction.

9 FAST FACT...

 MAPS ARE MADE at different scales for different purposes. A 1:25 000 scale map is very useful for walking but if you use it when driving, you'll quickly fall off the edge! A 1: 250 000 scale map shows much more land but in far less detail.

10 MAP READING — LEGEND OR KEY

JUST LIKE A KEY TO A DOOR, the key (also known as the legend) on a map helps you to unlock the information stored in the colors and symbols on a map. Colors and symbols are used to avoid too much information in the form of writing (there isn't room for that!). You must understand how the key relates to the map before you can unlock the information it contains.

The key is especially useful and will help you to identify everything from types of boundaries, roads, buildings, agriculture, industry, and places of interest to geographical features.

A series of easily recognizable symbols are commonly used to represent the different features. A place of worship for example, will typically be indicated by a symbol with a cross on it.

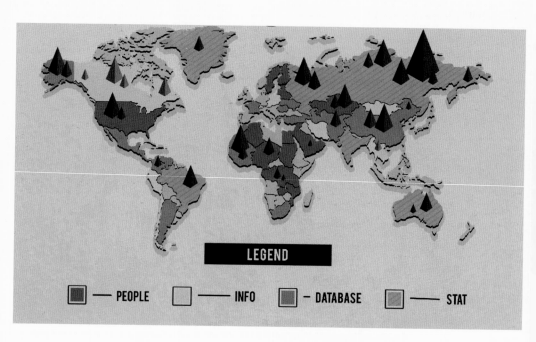

LEGEND

■ — PEOPLE   □ — INFO   ■ – DATABASE   ■ — STAT

11 MAP READING — TOPOGRAPHY

TOPOGRAPHY IS THE ART AND SCIENCE of depicting heights and depths on a map. Every location has an elevation with respect to sea level, and together they constitute 'the lay of the land', which is what topography is all about.

Topography is commonly represented in three different ways to give an indication of elevation.

Spot Heights – a symbol (typically a tiny dot, sign or triangle) accompanied by a number indicates the elevation (in feet or meters) of a given point.

Contour Lines – connect points of equal elevation and in so doing, convey the shape of the land they depict.

Shading – colors and gray tones may be used to indicate elevation above sea level, and refers to a range (rather than a single point) of elevations over an area. Remember to check the legend for the meaning of particular shades.

12 FAST FACT...

SPOT HEIGHT NUMBERING leads up hill, so the higher the number, the higher the elevation.

13 FAST FACT...

REMEMBER, the closer contour lines are together, the steeper the elevation, or the valley (contours can also indicate depressions).

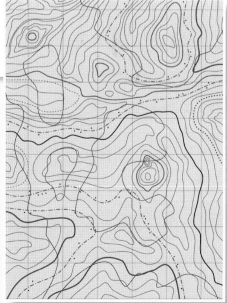

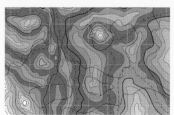

14 PHOTOGRAPHY AS A MAPPING TOOL

 A MAJORITY OF MAPS produced under government approval, from municipal to federal, are directly derived from aerial photography. That is, photos of the Earth's surface taken from aircraft, making it easier to identify and map surface features. Black and white film is generally used for this purpose as it's less expensive and provides a clearer view of Earth's surface than color.

Infrared photography is also a very popular mapping tool. Infrared energy, seen by special kinds of films and sensors, is contained in the sunlight that strikes the Earth and reflects off its surface. It can readily pass through air pollution, resulting in crisp images even on days when the atmosphere is far from clean. Infrared photography is thus widely used in aerial surveys.

15 FAST FACT...

INFRARED PHOTOGRAPHY is capable of providing information that may not otherwise be apparent. For example, shades of redness may even indicate plant life that is stressed due to disease or drought.

GPS (GLOBAL POSITIONING SYSTEM) MAPPING

IT IS OF COURSE IMPORTANT to get the positional accuracy of mapped objects, spot on. Historically, this would have meant field observation where an explorer or surveyor for example, would travel to a particular area in order to observe and map a location. Today though, positional accuracy can be provided through GPS (Global Positioning System), which is a space-based satellite navigation system providing location and time information in all weather conditions.

Providing latitude and longitude information, and sometimes altitude, although this is not considered sufficiently accurate or continuously available enough, GPS is used in military, aviation, marine and consumer product applications all over the world. GPS is freely accessible to anyone with a GPS receiver including for example a satellite navigation system or even a mobile phone.

GPS on cellphones

17 FAST FACT...

A GPS POSITION can be given anywhere on or near the Earth where there is an unobstructed line of sight to four or more GPS satellites.

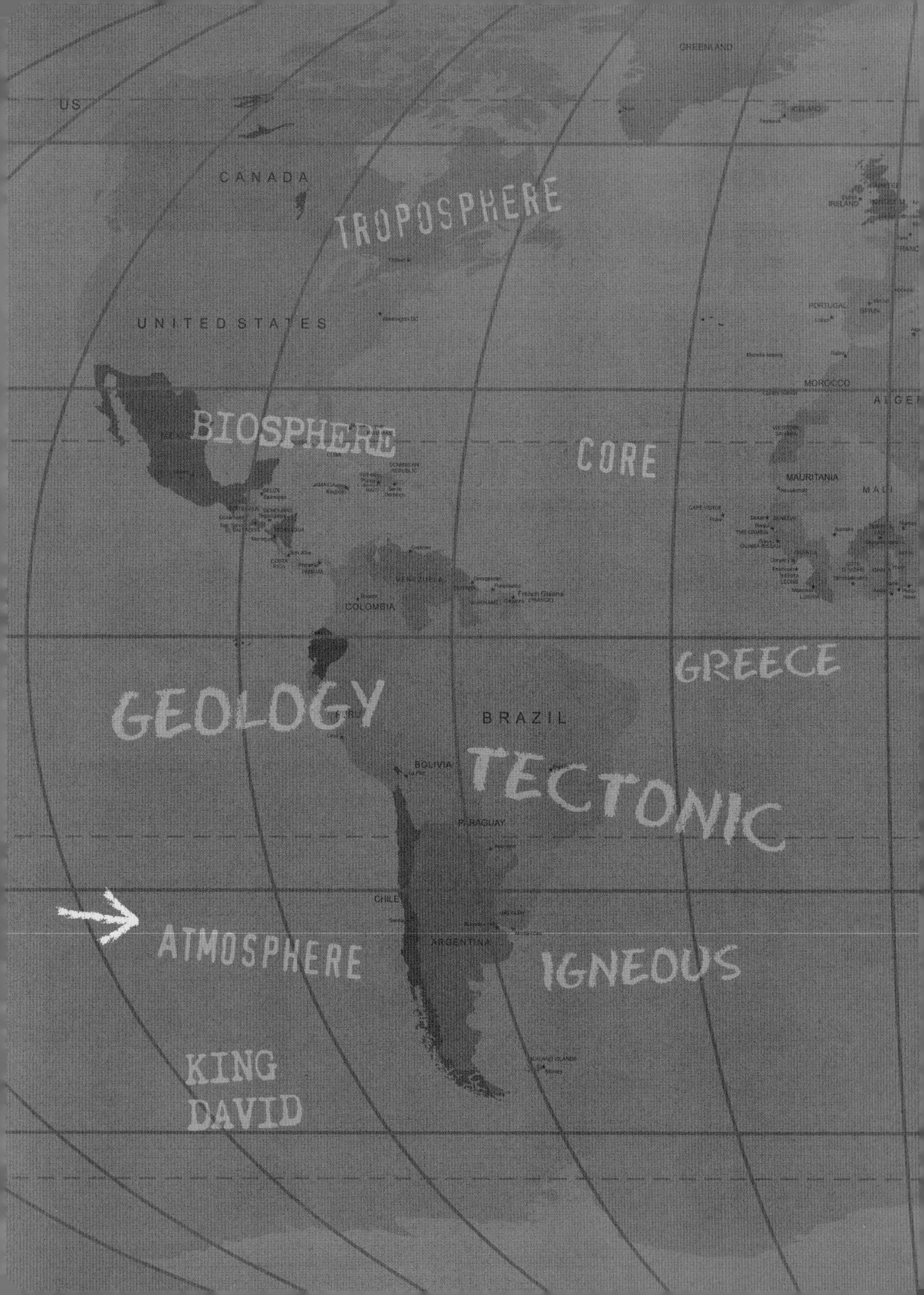
GREENLAND
US
CANADA
TROPOSPHERE
UNITED STATES
BIOSPHERE
CORE
MOROCCO
MAURITANIA
COLOMBIA
GREECE
GEOLOGY
BRAZIL
BOLIVIA
TECTONIC
PARAGUAY
CHILE
ATMOSPHERE
ARGENTINA
IGNEOUS
KING
DAVID

18 EARTH — WHERE WE LIVE

LUCKILY FOR US, conditions for life on Earth are just right. In fact, the combination of conditions is what makes life on Earth thrive.

The Earth is the third planet in our solar system and is 93 million miles (149 million kms) from the Sun. If it were closer, Earth would be too hot for life to survive and if it were much farther away, it would be way too cold.

The water on our planet also makes the Earth just right for life. Liquid water lets molecules easily come into contact, which helps chemical reactions needed for life.

Water also makes up an important part of living things. Even about 90% of our own bodies is water.

The Earth, from the surface right up into space, also has the right atmosphere for life. By volume, air is about 77% nitrogen and 21% oxygen. We and many other animals need oxygen of course, to breathe. The atmosphere acts a bit like a blanket capturing just enough of the Sun's warmth and holding it in. Without the atmosphere, Earth's sunny side would be way too hot, and its dark side would be bitterly cold. The atmosphere also acts as a sunscreen with its ozone layer screening out many of the Sun's harmful ultraviolet rays.

LAND AREA 29,1%

WATER AREA 70,9%

19 FAST FACT...

The **EARTH** was created around 5.5 billion years ago, and it's thought that life on Earth began shortly after this.

20 EARTH'S CRUST

👨‍🎓 **THE CRUST FORMS** the outer surface of the Earth. Originating about 4.7 billion years ago as a molten fireball, the Earth has slowly been cooling ever since. As a result, and after so many years, the outermost portion has hardened into a layer of rock and has cooled down.

Tectonic forces have built up the Earth's crust and have created the different kinds of landforms we see today – mountains, valleys, plains, plateaus, etc. Between 5 and 40 miles deep, which in reality accounts for a very small portion of planet Earth, the crust, is hugely important because, of course, it's home to life on Earth.

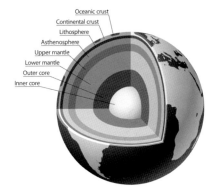

Oceanic crust
Continental crust
Lithosphere
Asthenosphere
Upper mantle
Lower mantle
Outer core
Inner core

Layers of the Earth

The Earth's crust, which is cool enough to behave much like a rigid shell floats on hotter, semi-molten rock known as mantle, and is subdivided into Continental and Oceanic plates.

There are seven large plates (for e.g., Pacific plate and Eurasian plate) and 12 smaller ones (for e.g., Phillipine plate and Juan de Fuca plate). The continental plates are mainly composed of granite and the oceanic plates, composed of basaltic rock are denser and younger, and are often destroyed beneath the continental plates, which float on top.

21 FAST FACT...

📖 **WHILE TECTONIC** forces build up the Earth's crust, Gradational Forces wear it down. The Grand Canyon is a result of gradational force.

22 DEEP WITHIN THE EARTH

BEYOND THE EARTH'S CRUST is a whole other World. The average distance from Earth's surface to its center is 3960 miles and no human has ever come close to seeing first-hand what it's really like deep within the Earth. In fact, we've barely penetrated the crust. Instead our understanding of Earth's interior rests on a combination of inference, analysis of alien objects (i.e., meteorites), sound waves, rocks, and minerals.

Directly beneath the lithosphere (which is the top layer, incorporating the crust) lies the asthenosphere. Measured in thousands of degrees Fahrenheit, its rock assumes a plastic, almost molten quality. Beneath the asthenosphere is a vast volume of somewhat stronger rock and below that liquid iron of the outer core and solid iron of the inner core, which is hotter than we could ever imagine.

23 FAST FACT...

THE POWERFUL heat of Earth's inner core is the source of pressure behind tectonic force.

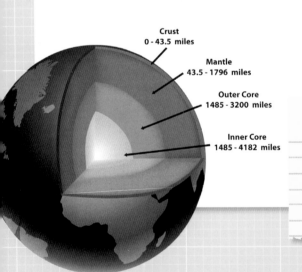

Crust
0 - 43.5 miles

Mantle
43.5 - 1796 miles

Outer Core
1485 - 3200 miles

Inner Core
1485 - 4182 miles

24 FAST FACT...

THE DISTANCE from the North Pole to the South Pole through the center of the Earth is 7,900 miles (12,714 kms).

25 4 INTERCONNECTED SPHERES

THE AREA NEAR THE SURFACE of the Earth can be divided up into four inter-connected geo-spheres; the lithosphere, hydrosphere, biosphere, and atmosphere. Scientists can classify life and material on or near the surface of the Earth to be in any of these four spheres.

Lithosphere – the solid, rocky crust composed of minerals that covers the entire planet from the top of Mount Everest to the bottom of the Mariana Trench.

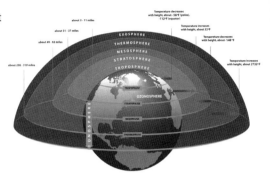

Hydrosphere – composed of all of the water on or near the Earth, it includes oceans (97% of Earth's water is here), rivers, lakes, and even the moisture in the air.

Biosphere – composed of all living organisms such as plants, animals, and one-celled organisms. Most of the planet's life is found from three meters below the ground to 30 meters above it and in the top 200 meters of the oceans and seas.

Atmosphere – the body of air which surrounds our planet, most of which is located close to the Earth's surface where it is most dense.

26 FAST FACT...

ALL FOUR SPHERES can be present in a single place. For example, a piece of soil may have mineral material from the lithosphere, moisture from the hydrosphere, insects or plants from the biosphere and even pockets of air from the atmosphere.

27 FAST FACT...

THE NAMES of the four spheres are derived from the Greek words for stone (litho), air (atmo), water (hydro), and life (bio).

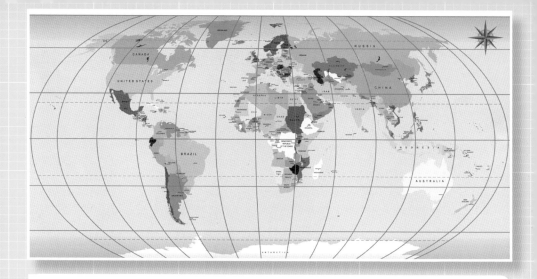

28 EQUATOR – LONGEST LINE OF LATITUDE

🎓 **THREE OF THE MOST SIGNIFICANT IMAGINARY LINES** running across the surface of the Earth are the Equator, the Tropic of Cancer, and the Tropic of Capricorn. While the Equator is the longest line of latitude on the Earth (the line where the Earth is widest in an east-west direction), the tropics are based on the Sun's position in relation to the Earth at two points of the year. All three lines of latitude are significant in their relationship between the Earth and the Sun.

The Equator is located at zero degrees latitude and divides the planet into the Northern and Southern Hemispheres. It runs through Indonesia, Ecuador, northern Brazil, the Democratic Republic of the Congo, and Kenya, among other countries, at which point the length of day and night are equal everyday of the year - day is always 12 hours long and night is always 12 hours long.

On the Equator, the Sun is directly overhead at noon on the two equinoxes - near March and September 21.

29 FAST FACT...

📖 **THE DISTANCE**
around the Equator is
24,902 miles (40,075 kms).

30 POLAR OPPOSITES — NORTH AND SOUTH POLES

🎓 **THE NORTH POLE,** located in the Arctic Circle, is the most northerly point on our planet. At the North Pole and the area around it, there is no land at all, just hundreds of miles of ice.

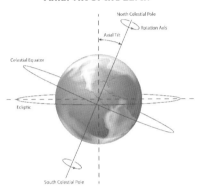

Axial Tilt of the Earth

North Celestial Pole
Rotation Axis
Axial Tilt
Celestial Equator
Ecliptic
South Celestial Pole

The South Pole, located in the Antarctic is the most southern point on the planet. Unlike the North Pole, the South Pole is located on land but it's buried under ice that's more than 1¼ miles (2 kms) thick.

Due to the tilt of the Earth's axis, the seasons are totally opposite at either pole. So, on a balmy summers day in the Arctic, when it might be as warm as 32°F (0°C), it's a teeth-chattering winters day in Antarctica, where the temperatures can reach as low as -56°F (-49°C).

Even though, the poles get plenty of sunshine in the summer months, it remains very cold and temperatures rarely climb above freezing. This is because the Sun is always low in the sky; its rays are weakened as they have to travel farther to reach the poles than they do to reach the Equator. The white ice, which covers the poles also reflects heat back into the atmosphere.

31 FAST FACT...

📖 **DUE TO GLOBAL WARMING,** scientists predict that by 2050, the Arctic ice sheet will be fragile enough for icebreakers to carve a straight path allowing ships to sail across the North pole.

32 ROCKS!

ROCKS FOUND ON THE EARTH'S SURFACE actually come from inside the Earth - so they tell us a lot about the Earth's interior. Based on their formation, they're organized into three groups:

Igneous rock – formed by magma from the molten interior of the Earth. They may either be of volcanic origin, forming when molten material, such as lava, is thrown out on to the surface of the Earth and cools quickly. Or, they may be formed when large masses of molten material are forced up into the crust and cool slowly under great pressure deep underground. This is known as intrusive rock, such as granite.

33 FAST FACT...

AN IMPERMEABLE ROCK allows very little water to pass through it whereas a permeable rock allows water to penetrate into it through pore spaces.

Metamorphic rock – have been subjected to tremendous heat and/or pressure, causing them to change into another type of rock. They are usually resistant to weathering and erosion and are therefore very hard-wearing. Examples of metamorphic rocks include marble, which originates from limestone and slate, which originates from clay.

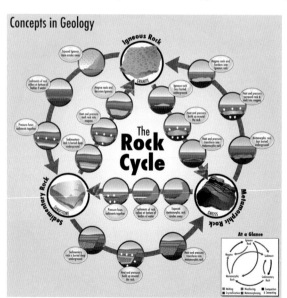

Sedimentary rock – forms from sediment that has settled at the bottom of a lake, sea or ocean, and have been compressed over millions of years. Examples of sedimentary rocks include sandstone, limestone, chalk, and clay.

34 SOIL — FOR PLANT GROWTH

BY DEFINITION, soil is a collection of Earth particles that are no more than 2 mms in diameter.

The soil that lies on Earth is a complex environment of minerals, organic material, water, gasses, and living organisms providing nutrients to plants, without which they simply would not thrive.

A fertile soil is one that makes lots of nutrients available to plants.

36 FAST FACT...

THE PROCESS by which nutrients are transferred from soil to the roots of plants is known as osmosis.

Various factors contribute to the production of soil, which averages out to be approximately ½ minerals, ½ water, and air. Overall, only a tiny proportion of soil consists of organic materials. The minerals that compose soil are formed by the erosion of rocks.

Soil is more prevalent in regions of high moisture and high temperatures than in cold, dry regions as the increased moisture contributes to erosion and increased temperature contributes to a more rapid breakdown of organic material. When organic material breaks down, it forms 'humus' – a dark tarry substance that is especially fertile for plant growth.

35 FAST FACT...

SOIL FERTILITY varies geographically and is one of several elements that explains why the population of one country may be better nourished than another.

37 WEATHERING THE PLANET

WEATHERING REFERS to the natural processes that break rock into smaller and smaller pieces. It can be distinguished from erosion because it need not involve any movement of material although weathering loosens material which can subsequently be transported by agents of erosion such as running water, wind, etc.

Weathering is broken down into two different categories.

Mechanical Weathering is the disintegration of rock and solid Earth material and its mechanisms include Frost Shattering or Free-thaw Weathering (where rocks allow water to occupy space between particles, and if the water freezes, the resulting ice crystals exert outward pressure that cracks) and Hot/Cold fluctuation (where rocks that are exposed to extreme fluctuations in temperature, may undergo expansion and contraction, resulting in some parts of rock breaking away).

38 FAST FACT...

WEATHERING WON'T TURN a mountain into a molehill but it does make possible the movement of surface materials by converting large immobile pieces of rock into small transportable ones.

Chemical Weathering is the disintegration of Earth material by chemical means such as oxidation, sometimes known as rusting. This occurs when rocks are exposed to oxygen in the air or water, which causes them to crumble more easily. Dissolution is another form of chemical weathering where direct and prolonged contact between water and certain rock minerals may cause the latter to decompose.

🎓 **GEOLOGY,** which is closely related to Geography, is the study of the Earth. It's thought that Earth is 4,600 million years old although geologists mostly focus on the last 600 million years!

Through the study of rocks, structures, processes, and fossils, geologists have established a geological timescale. This provides us with a common set of names and a timescale, which helps when describing and understanding different landscapes as well as particular major events that are thought to have happened in that time.

The most recent period in geological time is called the quaternary, when the Ice Age occurred. Before that was the Neogene period and before that the Palaeogene period.

40 FAST FACT...

📖 **THE JURASSIC PERIOD,** beginning after the Triassic period around 206 million years ago, is well-known as the age of the dinosaurs.

Rocks are formed during different periods and are a result of the environment present during that geological time. For example, the Cretaceous period saw the deposition of chalk.

WHY IS THE ATMOSPHERE SO IMPORTANT?

THE EARTH IS SURROUNDED BY THE ATMOSPHERE, which is the body of air or gases that protects the planet and enables life. A collection of five distinct layers (troposhere, stratosphere, mesosphere, thermosphere and exosphere), and several intermediate layers, the atmosphere starts at ground level, is measured at sea level, and rises into what we call outer space. The atmosphere has a very important protective role. The moon for example, which has no atmosphere, is covered with craters from being bombarded by meteors. Those meteors that might have hit the Earth however, burn up in the atmosphere before reaching us and all we see is the streak of light, known as a shooting star, they burn up.

42 FAST FACT...

📖 **BETWEEN EACH LAYER** of the atmosphere is a boundary, known as a 'pause'. At these 'pauses', maximum change between each layer occurs.

43 FAST FACT...

📖 **THE AIR OF OUR PLANET** is 79% nitrogen and just under 21% oxygen; the small amount remaining is composed of carbon dioxide and other gasses.

The atmosphere also protects us from the fierce ultraviolet light from the Sun. It absorbs much of this light and lets just enough through to warm and nourish us whilst also keeping us from getting too hot or too cold. You could say that life on Earth evolved the way it did because our atmosphere provided the exact conditions we need to live.

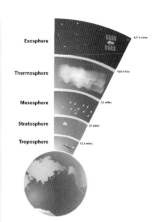

44 UP IN THE STRATOSPHERE

 **THE LAYER OF THE** atmosphere closest to the earth is the troposphere. This is where weather occurs. Beyond this though, is the stratosphere, which extends to about 30-35 miles above the Earth's surface. Temperatures rise within the stratosphere but still remain well below freezing.

Here, the air doesn't flow up and down, but flows parallel to the Earth in very fast moving air streams. It's because of these air streams and the fact that it's a much calmer environment than in the weather-dependent troposphere that most airplanes fly in the stratosphere.

The top edge of the stratosphere is abundant with ozone. Ozone is the byproduct of Sun radiation and oxygen. By capturing the ultraviolet rays of the Sun and deploying it, ozone takes out the harmful effects. This is very important to all living things on Earth, since unfiltered radiation from the Sun can destroy all animal tissue.

45 FAST FACT...

THE STRATOSPHERE is where ultraviolet radiation from the Sun reacts with oxygen to form ozone gas and the ozone layer, which protects us from the Sun's harmful radiation.

46 FAST FACT...

BETWEEN EACH LAYER of the atmosphere is a boundary, known as a 'pause', where maximum change between the spheres occurs.

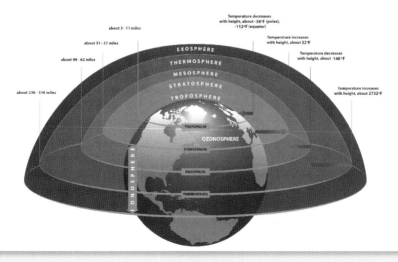

about 3 - 11 miles

about 31 - 37 miles

about 49 - 62 miles

about 236 - 310 miles

Temperature decreases with height, about -58°F (poles), -112°F (equator)

Temperature increases with height, about 32°F

Temperature decreases with height, about -148°F

Temperature increases with height, about 2732°F

EXOSPHERE
THERMOSPHERE
MESOSPHERE
STRATOSPHERE
TROPOSPHERE

TROPOPAUSE
OZONOSPHERE
STRATOPAUSE
MESOPAUSE
THERMOPAUSE

IONOSPHERE

47 TROPOSPHERE — MEANING 'WHERE THE AIR TURNS OVER'

 THE LAYER OF THE ATMOSPHERE CLOSEST TO THE EARTH is called the troposphere. It begins at the surface of the Earth and extends out to about 4-12 miles. This layer is where all weather occurs as the air in the troposphere is in a constant up and down flow. The temperature of the troposphere decreases with height as it gets closer to the Stratosphere.

The troposphere is immediately important in our daily activities. The bottom third, that which is closest to us, contains 50% of all atmospheric gases and this is the only part of the whole makeup of the atmosphere that is breathable. You could say then that the Troposphere keeps us alive.

The troposphere also has a north–south oriented aspect. The air from the northern hemisphere and the air from the southern hemisphere meet and mingle at the Equator, but never go any farther.

48 FAST FACT...

THE TROPOSPHERE is where all weather – including clouds, lightning, high winds, hurricanes, tornadoes, snow, hail and freezing rain – forms and occurs.

49 SUNRISE AND SUNSET

AS THE SUN RISES, it peeks above the eastern horizon and as it does so, changes from reds into shades of orange and yellow, bringing with it a bright new day. Just as awe-inspiring is the sparkling palette that sunset paints across the sky. While the eastern sky may still be deep blue, streaks of orange, pink, and red, light the western sky as the Sun sinks lower below the horizon.

But, what causes the Sun to come up and go down each day? Well, although the Sun appears to move, it's really Earth that's moving; always spinning or rotating, toward the East.

How about the beautiful colors associated with sunrise and sunset? During these times the Sun's rays must travel through more of the atmosphere to reach us. As light waves enter the atmosphere, they strike gas and particle molecules in the air. The molecules scatter the light. That is, they split the different colors apart and the shorter blue-toned wavelengths get scattered about four times more easily than the longer red-colored wavelengths, which, during sunrise and sunset, make the sky glow an orange-red color.

50 FAST FACT...

ALTHOUGH we usually can't see it, a phenomena known as the, 'green flash' sometimes happens literally, in a flash, just before sunrise or just after sunset when light's reddish rays may be hidden below the horizon, while the blue hues get scattered by the air.

51 FAST FACT...

NEVER LOOK DIRECTLY at the Sun – no matter what time of the day it is, harmful ultraviolet rays can damage your eyes.

52 I CAN SEE A RAINBOW... BUT HOW DID IT GET THERE?

IF YOU HAVE A PRISM, you can separate sunlight into its individual colors. A typical prism looks like a triangular column of glass and as light passes through, the glass bends the light. The bending lets you see the visible spectrum: red, orange, yellow, green, blue, indigo, and violet.

Rainbows result from nature's prisms. Sometimes after it stops raining overhead, droplets stay in the air ahead of you. If Sun shines from behind you through the droplets, the drops act like prisms.

But why do rainbows have an arch-like shape? Well, light's red wavelengths are longer than the blue ones therefore they arch over the shorter wavelengths thus making a rainbow curve.

53 FAST FACT...

'RAYLEIGH SCATTERING' is the process by which molecules in the atmosphere split or scatter different colors within white light.

54 FAST FACT...

DON'T SEARCH FOR a pot of gold at the end of the rainbow! As you move, the projection of the rainbow moves too and when the air dries, the rainbow disappears.

55 WATERY PLANET

WITH NEARLY 70% OF OUR PLANET'S SURFACE covered by water, it comes as no surprise that water is vital to life on Earth. From the world's five oceans to the many seas, lakes and rivers, water really is everywhere. In addition, water is in the air we breathe and is mixed up in the soil underfoot.

The geography of water is very unevenly distributed though and as such, determines any number of things, from where people live, where travel is permitted and where fishing is allowed, to the content of the atmosphere and even the shape of the world.

So, where did the water come from? Apparently it was here from the beginning. When Earth was a newborn fireball, its water was mixed together with other planetary matter. Back then, chances are that the majority of water was on the inside of the planet rather than on its surface. Over time it migrated to the outside until the Earth's crust cooled.

56 FAST FACT...

WATER may cover almost 70% of Earth's surface but it accounts for just 0.5% of the planet's weight.

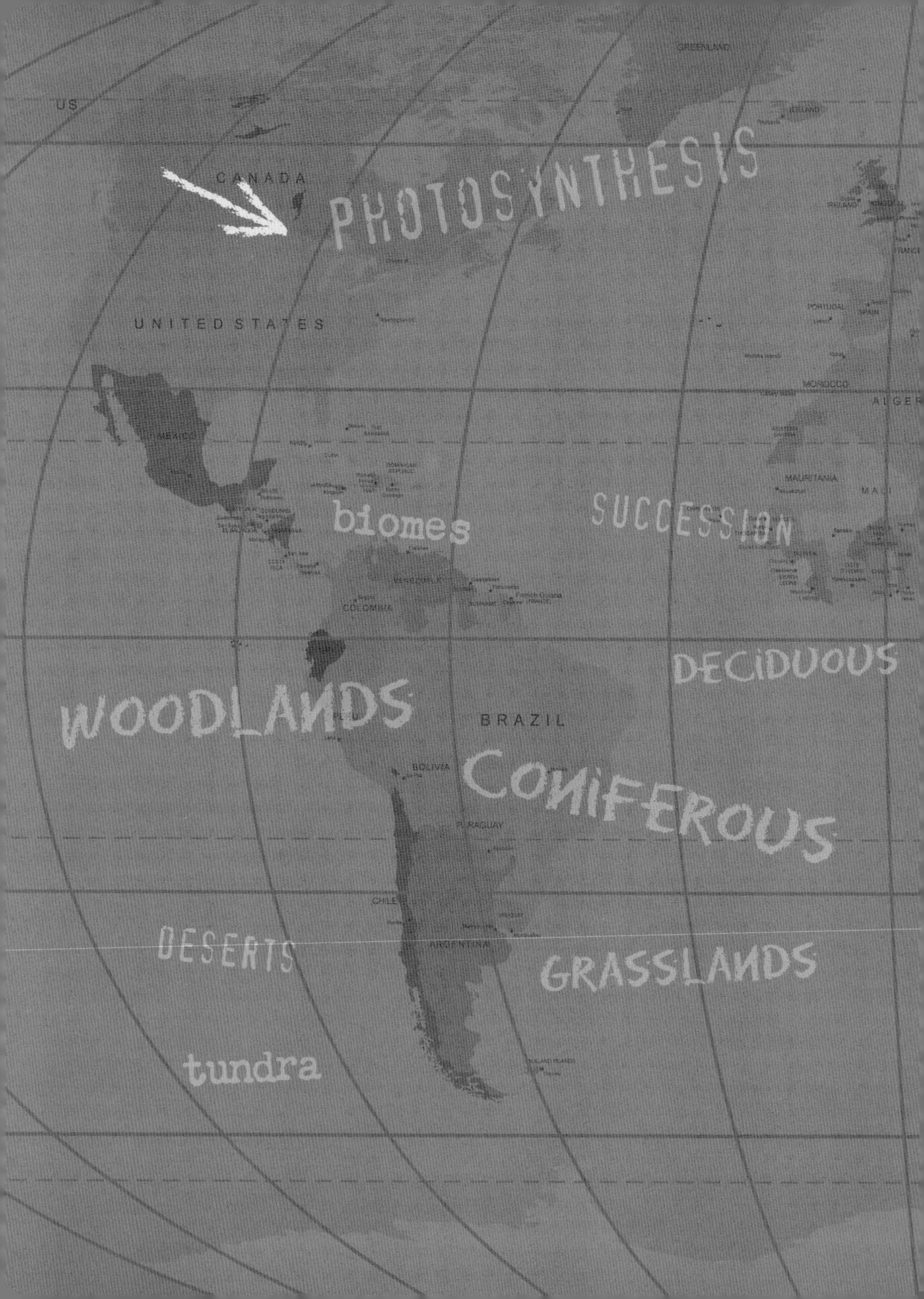
PHOTOSYNTHESIS
biomes
SUCCESSION
DECIDUOUS
WOODLANDS
CONIFEROUS
DESERTS
GRASSLANDS
tundra
US
GREENLAND
ICELAND
CANADA
IRELAND
UNITED STATES
PORTUGAL
SPAIN
MOROCCO
MEXICO
MAURITANIA
MALI
VENEZUELA
COLOMBIA
French Guiana (FRANCE)
BRAZIL
BOLIVIA
PARAGUAY
CHILE
ARGENTINA

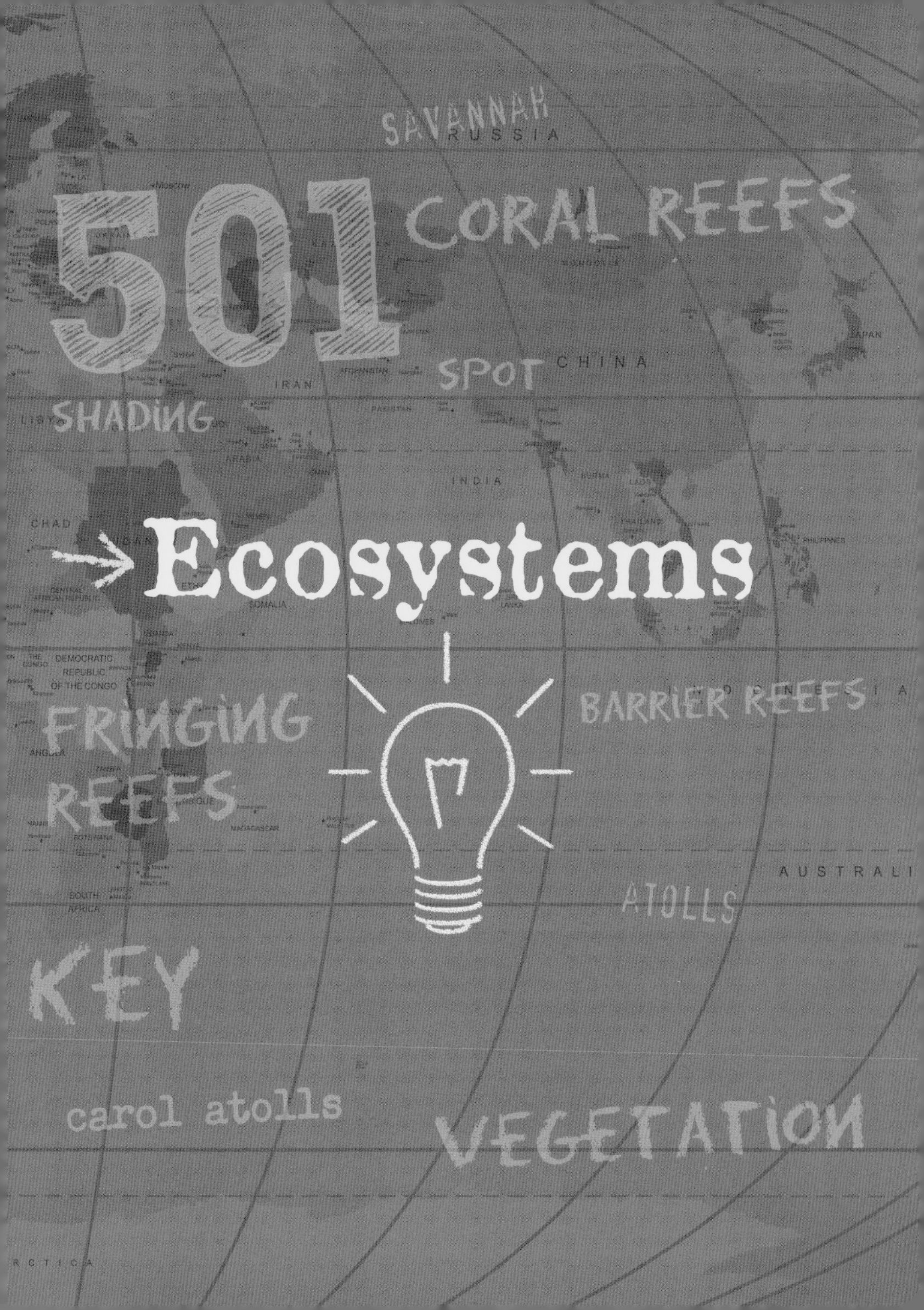

SAVANNAH
RUSSIA

501 CORAL REEFS

SPOT

SHADING

→ **Ecosystems**

FRINGING
REEFS

BARRIER REEFS

ATOLLS

KEY

carol atolls

VEGETATION

57 UNDERSTANDING ECOSYSTEMS

🎓 **AN ECOSYSTEM DESCRIBES** the links between the living community of animals and plants and their habitat.

Ecosystems can be small-scale, i.e., covering a small area such as a pond or an oak tree, or large-scale covering a large area such as a tropical rainforest or coral reef. The large scale ecosystems are known as biomes and key to every biome is the energy cycle, the nutrient cycle and a process known as succession where events follow in order of sequence.

58 FAST FACT...

📖 **THE WORLD IS DIVIDED** up into 10 major ecosystems. These large-scale ecosystems are called biomes.

Energy Cycle – the Sun is the source of all energy in an ecosystem. Energy flows through the system as a food web, i.e., plants absorb the Sun's energy before being eaten by animals, and animals eat each other!

Nutrient Cycle – weathered rock releases nutrients such as calcium, carbon, and nitrogen into the soil. Plants and animals then circulate nutrients through the system. Nutrients can be lost into rivers.

Succession – the process by which ecosystems develop over time. For example, more nutrients are supplied through weathering and decomposition of plants allowing new plants to germinate. Given time, a dominant species takes over and smaller plants live beneath the dominant one.

59 FAST FACT...

📖 **PHOTOSYNTHESIS** is the name given to the process whereby plants absorb light energy from the Sun.

60 TROPICAL RAINFOREST

A TROPICAL RAINFOREST IS A BIOME FOUND IN HOT, humid environments in equatorial climates. The level of humidity and density of the vegetation give the ecosystem a unique water and nutrient cycle causing a rainforest to have the most diverse range and highest volume of plant and animal life found anywhere on Earth.

There are distinct layers of vegetation with the tallest trees in the rainforest up to 50 meters tall. These giants have surprisingly shallow roots but develop buttress roots above ground to support their height. At ground level, the forest is dark as little sunlight can penetrate the dense canopy.

61 FAST FACT...

THE AMAZON rainforest has approximately 300 species of tree (mostly hardwoods like ebony, teak and mahogany) per square mile.

It's no coincidence they they're called rainforests as, due to the high humidity and rising air, it rains virtually everyday, although the level of rainfall depends on the time of year. With so many trees and such a dense canopy though, nearly all the rain is intercepted.

62 FAST FACT...

RAINFORESTS around the world are threatened by human expansion.

63 CONIFEROUS WOODLANDS

 CONIFEROUS OR BOREAL FORESTS OCCUR IN COLD NORTHERN REGIONS (found between 50° and 60° north of the Equator) characterized by long cold winters and short summers. Although there is a lot of precipitation here, most of it falls as snow.

Coniferous woodlands are characterized by evergreen coniferous trees like spruce, pine and fir, with needles instead of leaves. These trees have thick bark to protect against the cold and are cone-shaped, with flexible branches which help them to cope with heavy snow fall.

Further north, the trees are shorter and less dense because of the shorter growing season but at its southern margins, the trees become taller and denser and merge with deciduous trees (those that lose their leaves).

The typical soil found in coniferous woodlands is a thin, ash-gray acidic soil with little organic content, known as a podsol.

64 FAST FACT...

📖 **ANIMALS COMMONLY** found in the Coniferous woodland region are reindeer, voles and hares.

65 DECIDUOUS WOODLANDS

CONTAINING TREES WITH BROAD LEAVES SUCH AS OAK, beech and elm, deciduous woodlands occur in places (between 40° and 60° north and south of the Equator) with high rainfall, warm summers and cooler winters.

The characteristic broad leaves form the canopy layer in summer but in autumn and winter of course, the trees loose their leaves. Some light is then allowed to get through and beneath the canopy the vegetation is layered. Beneath the taller trees is a shrub layer containing species like hazel, ash and holly whilst grass, bracken or bluebells can be found in the ground layer.

As the leaves fall off the trees, they decompose and help to give the soil its nutrients, thus the soil is a fertile brown Earth soil. Earthworms in the soil help to mix the nutrients, and blend the layers within the soil.

66 FAST FACT...

DECIDUOUS WOODLANDS have deep tree roots and so help to break up the rock below, which in turn helps to give the soil more minerals.

67 FAST FACT...

SOME OF THE MAJOR deciduous forests are in southwest Russia, Japan, and eastern China.

68 SAVANNAH GRASSLANDS

THE SAVANNAH BIOME is located further away from the Equator than the tropical rainforest biome in the central part of Africa and in South America. The Savannah is dry, but not as dry as desert areas.

There are two distinct seasons in the Savannah grasslands - a wet season and a dry season. In the wet season vegetation grows, including lush green grasses and wooded areas but moving further away from the Equator and its heavy rainfall, the grassland becomes drier and drier - particularly in the dry season when there is very little rain at all.

69 FAST FACT...

THE LARGEST expanses of Savannah are in Africa, where much of the central part of the continent, for example Kenya and Tanzania, consists of tropical grassland.

Savannah vegetation includes scrub, grasses and occasional trees, which grow near water holes, seasonal rivers or aquifers. Plants have to adapt to the long dry periods, e.g., the acacia tree with its small, waxy leaves and thorns. Plants may also store water, (e.g., the baobab tree) or have long roots that reach down to the water table.

Animals, of which there are many, including giraffes, zebras, elephants, lions, wildebeest, etc., also have to adapt and may migrate great distances in search of food and water.

70 FAST FACT...

MANY OF THE ANIMALS found on the Serengeti plains, including over 2 million wildebeest, can be found nowhere else in the world.

DESERTS

🎓 **THE EXTREME CLIMATE OF DESERTS** – hot during the day and cold at night with very little rainfall throughout the year - is what defines them.

During the day temperatures may reach 122°F, when at night they may fall to below 32°F. Most deserts are found between 20° and 35° north and south of the Equator, which are zones of tropical high pressure where hot descending air prohibits cloud development. The absence of cloud cover during the day permits the strong solar heating and rapid cooling at night.

72 FAST FACT...

📖 **THE SAHARA** is the largest desert, covering 3,500,000 m². It also has the hottest recorded temperature, at 136.4°F.

Deserts have less than 9.84 inches of rainfall per year and rain is very unpredictable, often falling in brief sudden downpours, causing flash floods. Even in a flood situation though, evaporation quickly exceeds rainfall.

Deserts are of course, extremely arid but animals (especially reptiles and insects) and plants (mostly low-growing shrubs and trees) have adapted to survive in the harsh conditions.

73 TUNDRA

TUNDRA BIOMES, found mostly along and within the Arctic Circle (to include Alaska, northern Canada, northern Europe, Russia and Siberia) are cold with very little precipitation. The climatic conditions mean, unsurprisingly, that the landscape is quite bare with no trees. In fact, the word 'tundra' means 'barren land'.

The temperatures stay below 32°F most of the year and the ground remains frozen, apart from a few centimeters of thaw in the summer. The precipitation is gentle, mainly falling as snow, and the winds can be very strong. Summers may have many hours of continuous daylight but winters are long, dark periods.

Vegetation has adapted to these conditions and exists mainly in the form of low growing mosses, lichens, hardy grasses, and dwarf shrubs. Animals meanwhile, are limited with only small herbivores such as lemmings and hares, remaining all year together with hunters such as arctic foxes and owls.

74 FAST FACT...

MOST ANIMALS in the tundra are summer visitors only. For example, reindeer migrate south into the coniferous forest in the winter.

75 FAST FACT...

WHERE TUNDRA landscapes have been exploited for minerals, such as oil and natural gas, the development of infrastructure has been difficult and often destructive to the environment.

76 SAND DUNES

SAND DUNES ARE KNOWN AS SMALL-SCALE ECOSYSTEMS. They form along coasts where there is a large supply of sand and strong prevailing winds blowing in from the sea.

Although small, they are fascinating features. As sand is blown inland, low dunes are formed, which are colonized by marram grass. Further sand is trapped by the grass, which has helped stabilize the dune, as it grows higher. The grass has adapted to the dry and windy environment by establishing long roots to reach the water table and strong leaves that can stand exposure to high winds and prevent excessive moisture loss. The grass also grows quickly so that it's not smothered by fresh sand.

In time, shrubby plants such as sea buckthorn, hawthorn, and gorse develop, which are colonized by insects, birds, rabbits, birds of prey, and foxes. Pine trees and eventually oak and ash complete the succession.

77 FAST FACT...

SAND DUNES are a common feature on sand spits, which are deposition landforms, connected at one end to a headland and extending outwards.

78 FAST FACT...

WIND CAN SOMETIMES remove the sand from dunes, leaving blow-outs. This is when root systems are exposed, which kills grasses and in turn leads to greater exposure of bare sand.

79 CORAL REEFS

A CORAL REEF IS A MARINE BIOME, in which the habitat and the living community are broadly the same. Corals for example, come from the same family as jelly fish but they secrete hard skeletons of limestone that surround and protect their soft bodies. Reefs are formed by large colonies of corals that slowly build hard structures upwards on the remains of previous colonies. This forms a habitat for themselves and for a huge variety of other plants and animals.

Corals are found in many marine environments, but colonial corals, those which are responsible for reefs, are found only in tropical waters between 30°N and 30°S along Pacific, Atlantic, and Indian coastlines.

There are 3 main types of coral reef:

Fringing Reefs – grow in shallow waters separated from the coast by a narrow channel.

Barrier Reefs – larger and more continuous, they grow parallel to the coast but are further seaward on the continental shelf.

Carol atolls – often low, circular reefs enclosing a lagoon, are formed around a volcanic island that has subsided.

80 FAST FACT...

AT 1,600 MILES LONG, the Great Barrier Reef in Queensland, Australia is the largest coral reef in the world.

81 FAST FACT...

📖 **CORAL REEFS ARE**
considered by scientists to
be possibly the most diverse,
productive communities on Earth.

82 FAST FACT...

📖 **REEFS PROVIDE PROTECTION**
from coastal erosion by absorbing
much of the energy of storm surges
created by cyclones or tsunamis.

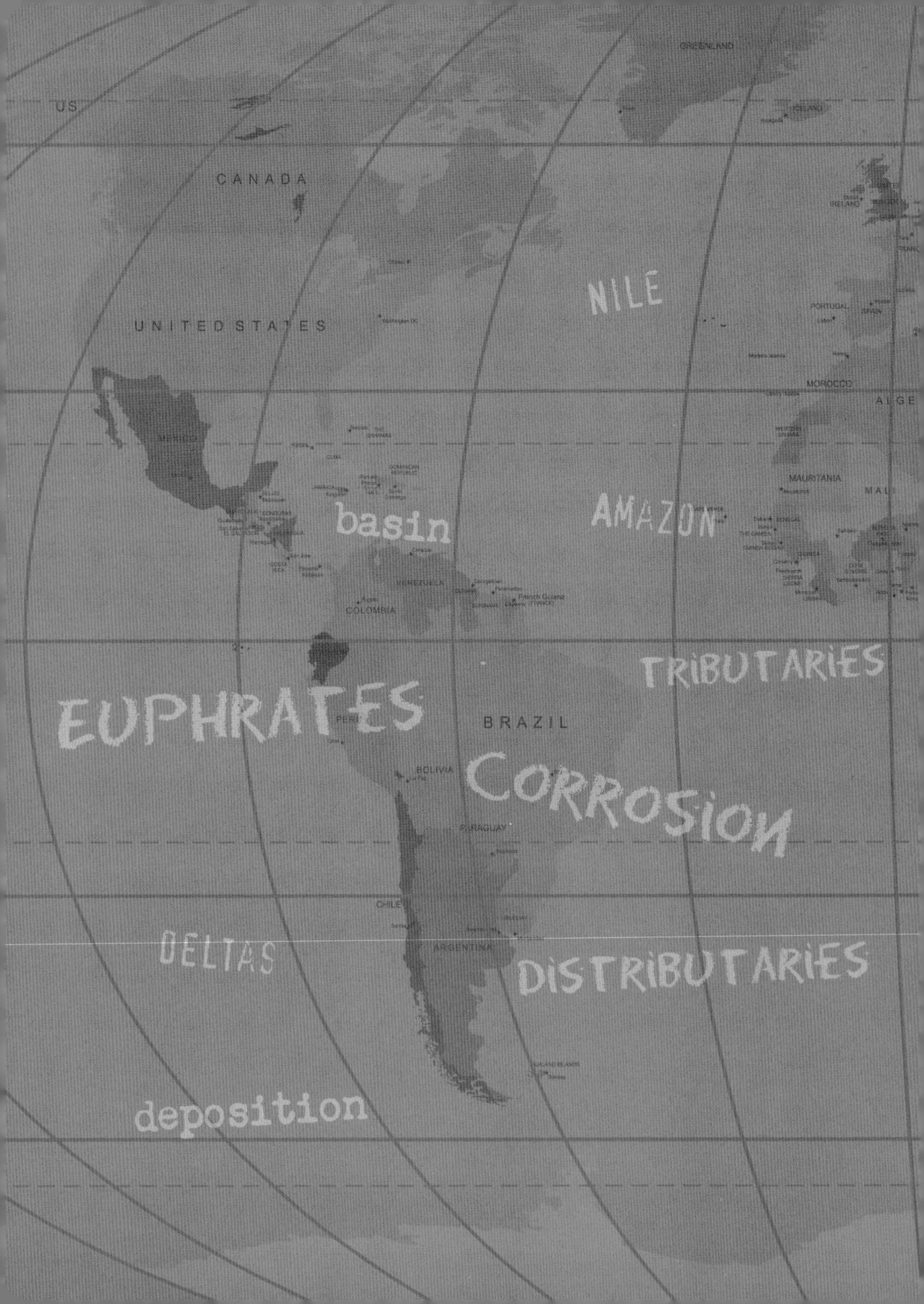

GREENLAND
ICELAND
US
CANADA
IRELAND
UNITED STATES
NILE
PORTUGAL
SPAIN
MOROCCO
MEXICO
MAURITANIA
basin
AMAZON
VENEZUELA
COLOMBIA
French Guiana
TRIBUTARIES
EUPHRATES
PERU
BRAZIL
BOLIVIA
CORROSION
PARAGUAY
CHILE
DELTAS
ARGENTINA
DISTRIBUTARIES
deposition

Rivers

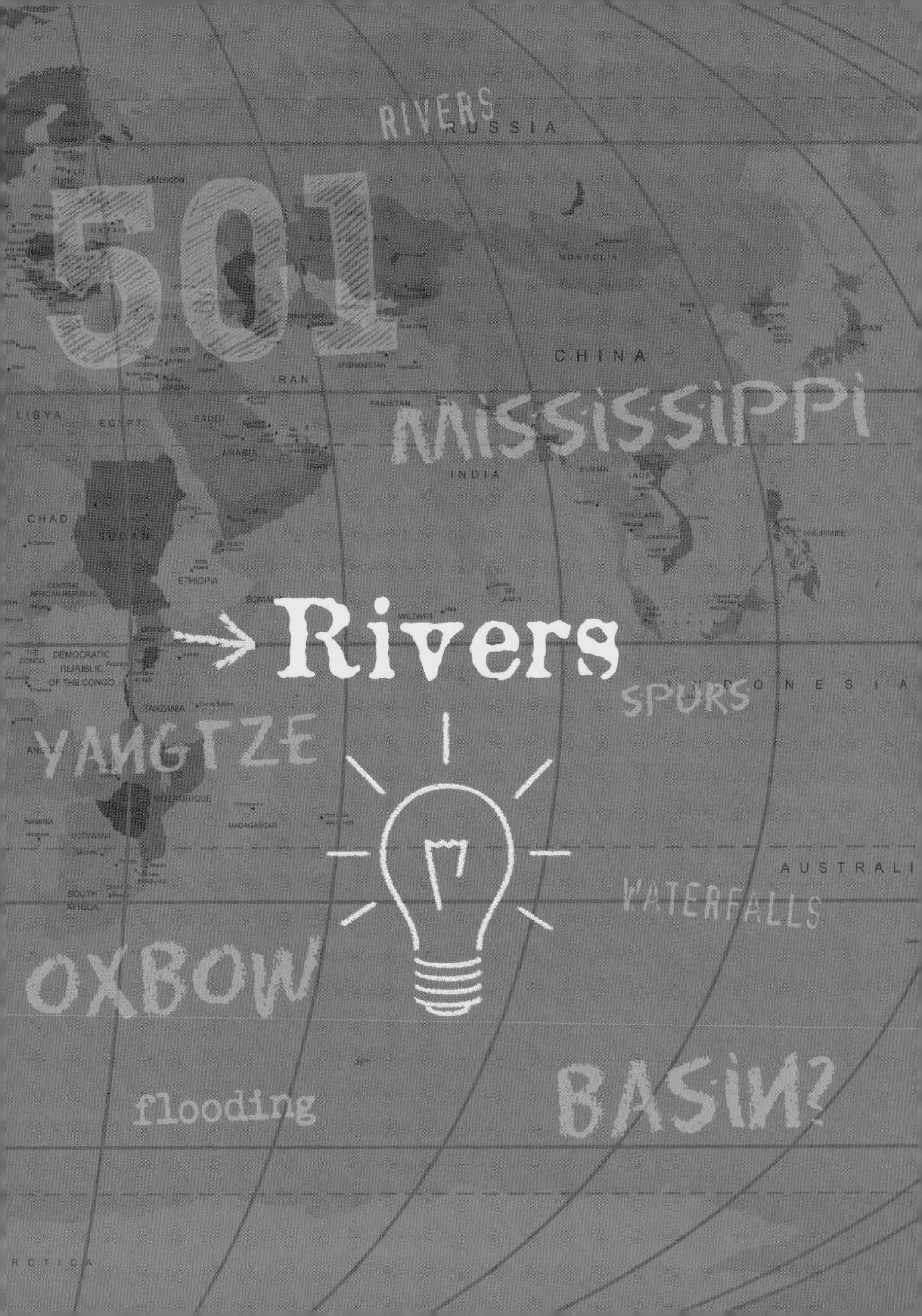

501

RIVERS

MISSISSIPPI

SPURS

YANGTZE

WATERFALLS

OXBOW

flooding

BASIN?

83 LONGEST!

🎓 THE FIVE LONGEST RIVERS IN THE WORLD ARE:

- River Nile (Africa) - 4,160 miles - 6,695 kms
- Amazon River (South America) - 4,049 miles - 6,516 kms
- Yangtze River (Asia) - 3,964 miles - 6,380 kms
- Mississippi-Missouri River System (North America) - 3,709 miles - 5,969 kms
  The Missouri River is, hydrologically, the upstream continuation of the Mississippi River as the Missouri River carries more water than the Mississippi River at the confluence of the two rivers.
- Ob-Irtysh Rivers – (Asia) - 3,459 miles - 5,568 kms

Longest of all, the River Nile is a major north-flowing, international river. Its water resources are shared by 10 countries, namely, Tanzania, Uganda, Rwanda, Burundi, Democratic Republic of the Congo, Kenya, Ethiopia, Eritrea, South Sudan, and the Arab Republic of Egypt. In particular, the Nile provides the primary water resource and so is the life artery for its downstream countries like Egypt and Sudan.

Before the building of a dam at Aswan, Egypt used to experience annual floods from the Nile that deposited 4 million tons of nutrient-rich sediment, thus enabling agricultural production. This process began millions of years before Egyptian civilization began in the Nile River valley and continued until the first dam at Aswan was built in 1889.

84 FAST FACT...

📖 COMPLETED IN 1970, after
10 years in the building, the Aswan High Dam (known as Saad el Aali in Arabic) captures the River Nile in the world's third largest reservoir, Lake Nasser.

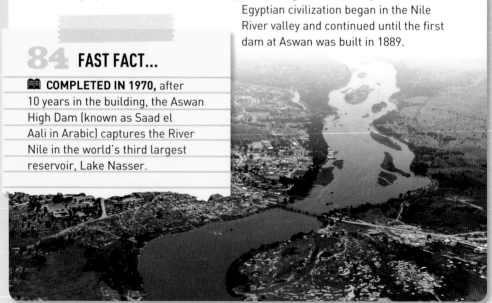

85 AMAZON RIVER — CARRYING MORE WATER THAN ANY OTHER RIVER

THE AMAZON RIVER CARRIES more water than any other river in the world. In fact, the Amazon River is responsible for about 20% of the fresh water that flows into the world's oceans.

At about 4,000 miles (6,400 kms), it's the second longest river in the World although ongoing studies are being carried out to substantiate some claims that it might in fact be longer than the River Nile.

86 FAST FACT...

THE AMAZON has taken different routes since it first began to carry water. Some scientists have determined that it even flowed west at one time or more, into the Pacific Ocean.

The Amazon River has the largest watershed (area of land that flows into the river) and more tributaries than any other river in the world. Streams that begin in the Andes Mountains are the starting sources for the Amazon and most of the run-off of Brazil along with runoff from Peru, Bolivia, Colombia and Ecuador flows into the Amazon.

For much of its path, the Amazon River can be as much as one to six miles wide. During flood seasons, the Amazon River can be much, much wider; some report it is more than 20 miles wide (32 kms) in certain places.

87 CROSS PROFILE – A SLICE ACROSS THE RIVER

THE CROSS PROFILE OF A RIVER is a slice across the river and a way of measuring the volume of water, known as discharge, in a cross-sectional area, multiplied by average velocity.

It might seem logical for a river to be flowing at its fastest at the source, in its upper course where the gradient is steep. It's not true though. Whilst waterfalls are almost certainly in the upper course if you look over a significant cross section of river, the fastest velocities will be found where the influence of the gradient is enough to defeat the dark forces of friction. In the end, it is in the lower courses, where the river channel is most efficient that the average velocity is at its highest.

88 FAST FACT...

THE UNIVERSALLY recognized symbol for river discharge is "Q."

To find the cross profile of a river, the average width of the river must first be found by measuring from bank to bank at various intervals along the river. Then the depth must be measured at regular intervals across the width of the river. The number of readings taken will depend on the width of the river.

THE LONG PROFILE OF A RIVER refers to the 'make-up' of a river, how it changes shape from source to mouth. The source of a river is often – but not always – in an upland area and flows over steep slopes with an uneven surface. It often flows over a series of waterfalls and rapids. Highland areas are usually composed of hard igneous rocks, which are ideal for forming such features.

As a river flows down steep slopes the water performs vertical erosion. This form of erosion cuts down towards the river bed and carves out steep-sided V-shaped valleys. As the river flows towards the mouth, the slopes become less steep. Eventually the river will flow over flat land as it approaches the sea and the discharge (amount of water flowing) will increase as the river approaches the sea.

90 TRIBUTARIES – FLOWING INTO THE MAIN STEM

🎓 **A TRIBUTARY IS A STREAM** or river which flows into the main stem of a river or lake, and not directly into the ocean.

91 FAST FACT...

📖 **THE AMAZON RIVER** has more tributaries (over 200 in total) than any other river in the world.

Tributaries are sometimes listed starting with those nearest to the source of the river and ending with those to those nearest to the mouth of the river. Sometimes though, they might be listed according to the Strahler Stream Order, which examines the arrangement of tributaries in a hierarchy of first, second, third, with the first order tributary being typically the least in size. A second order tributary would be the result of two or more first order tributaries (known as a confluence) combining to form the second order tributary, and so on.

Tributaries might also be referred to as 'right tributary' or 'left tributary', which are terms stating the orientation of the tributary relative to the flow of the main stem river. These terms are defined from the perspective of looking downstream.

92 FAST FACT...

📖 **THE WORLD'S LONGEST** and largest rivers can trace their origins to tributaries far, far away. The Euphrates River, which is the longest river in South West Asia for example, flowing through Syria and Iraq before draining into the Persian Gulf, is fed by tributaries originating in the mountains of Turkey.

93 DISTRIBUTARIES – WHERE A RIVER DIVIDES

🎓 **LESS-COMMON THAN A STREAM** or river flowing into the main stem, is where a river divides into smaller streams. This is known as bifurcation (from the Latin words for double and fork) and the resulting streams are known as distributaries.

There are three main types of river bifurcation. The first is when, after bifurcation, the streams merge again in one main river course. Some waterfalls are also short-period bifurcating objects like Niagara Falls where Goat Island divides the water stream. The second type of river bifurcation is when distributaries inflow in one drainage basin in a short distance. All coastal deltas are examples of this type. In this case the slow velocity of the river stream and the great amount of sediments cause bifurcation and create many alluvial islands. The third type is a rare natural phenomenon. The separated streams runoff into different drainage basins and they do not meet each other again.

94 FAST FACT...

📖 **THE GERMAN POLYMATH** and great geographer Alexander von Humboldt, together with the French botanist Aimé Jacques Alexandre Bonpland, were the first to discover and describe a distributary, found on the Amazon River in the Great Expedition.

95 DELTAS – THREE MAIN TYPES

DELTAS ARE FOUND AT THE MOUTH OF LARGE RIVERS such as the Mississippi or the River Nile and are so-called because many of them, most

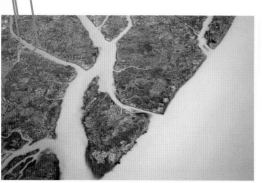

notably that of the River Nile, assume a roughly triangular shape reminiscent of the Greek letter delta.

A delta is formed when the river disgorges eroded sediment that's been carried along, faster than the coastal sea waters can remove it. Particles, mostly silt, are therefore deposited and progressively accumulate to form a delta.

There are three main types of delta, named after the shape they create:

Arcuate or fan-shaped - the land around the river mouth arches out into the sea and the river splits many times on the way to the sea, creating a fan effect.

Cuspate - the land around the mouth of the river juts out arrow-like into the sea.

Bird's foot - the river splits on the way to the sea, each part of the river juts out into the sea, rather like a bird's foot.

96 FAST FACT...

THE NILE DELTA in Northern Egypt is around 100 miles (160 kilometers) in length and spreads out over 149 miles (240 kilometers) of coastline. It is rich in agriculture and has been farmed for thousands of years. Today, around half of Egypt's population live in the Nile Delta region.

97 FAST FACT...

THE COLORADO RIVER delta, while once rich marshland, is today mainly dry aside from exceptionally wet years due to the removal of water upstream for irrigation and city uses.

98 DRAINAGE BASIN

🎓 **A DRAINAGE BASIN,** sometimes called a watershed, is an entire river system, or an area drained by a river and its tributaries. Drainage basins can cover very wide areas. The Amazon River drainage basin for example, is huge, draining over a third of the entire South American continent.

Some drainage basins are sharply defined by the crest of a high ridge, or by a continental divide. The Continental Divide of the Americas for example, roughly follows the crest of the Rocky Mountain range. Rain, snow and other precipitation falling on the west side of this divide flows into the Pacific Ocean. Precipitation falling on the east side flows into the Atlantic and Arctic Oceans.

99 FAST FACT...

📖 **SOMETIMES A DRAINAGE** basin flows into an internal body of water, called endorheic basins, such as the Dead Sea. Water can only leave endorheic basins by evaporating or seeping through the soil, which explains why the Dead Sea is one of the saltiest bodies of water on Earth.

One of the reasons drainage basins are important to scientists is that they affect the quality and amount of flow through a river at a given point. For example, as the Mississippi River empties into the Gulf of Mexico, it is carrying water from its entire watershed, the second-largest in the world. It includes about 40% of the area of the continental United States and provides water for millions of people.

100 RIVER PROCESSES

RIVER PROCESSES, or erosion, shape the land in different ways as the river moves from its source to its mouth, and can be easily remembered as CASH.

Corrasion – rocks rubbing against the bed and banks to change the channel shape.

Attrition – rocks rubbing against each other in the river to produce more rounded and smaller particles.

Solution – particles dissolving into the river. For example, limestone and atmospheric pollutants cause acidic water to dissolve particles along the river course.

Hydraulic Action – the force of the water against the bed or banks can cause air to become trapped, with the pressure weakening the bank and causing it to wear away.

101 FAST FACT...

EROSION may only occur in particular areas of fast-flowing water, like rapids or at certain times of the year when discharge is high.

102 TRANSPORT METHODS

RIVERS NEED ENERGY to transport material, and levels of energy change as the river moves from source to mouth.

When energy levels are very high, usually near the source of a river when its course is steep and its valley narrow, large rocks and boulders can be transported. When energy levels are low though, only small particles, if any, will be transported. Energy levels are lowest when velocity drops as a river enters a lake or sea (at the mouth).

There are four different river transport processes.
- Solution - minerals are dissolved in the water and carried along in solution.
- Suspension - fine light material is carried along in the water.
- Saltation - small pebbles and stones are bounced along the river bed.
- Traction - large boulders and rocks are rolled along the river bed.

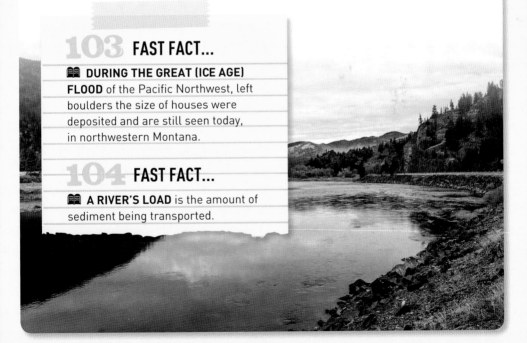

103 FAST FACT...

DURING THE GREAT (ICE AGE) FLOOD of the Pacific Northwest, left boulders the size of houses were deposited and are still seen today, in northwestern Montana.

104 FAST FACT...

A RIVER'S LOAD is the amount of sediment being transported.

105 DEPOSITION – SHAPING THE LANDSCAPE

 **WHEN A RIVER LOSES ENERGY,** it will drop or deposit some of the material it is carrying. Known as deposition, this happens most commonly when a river enters an area of shallow water or when the volume of water decreases, for example, after a flood or during times of drought.

Deposition is common towards the end of a river's journey, and when it occurs at the mouth of a river, it can form deltas.

Not just deltas, but deposition can cause other significant changes in the landscape. While the fringing slopes (making a V shape) of a river provide evidence of the continued presence of and power of erosion for example, the appearance of a flat and comparatively expansive valley floor (creating more of a U shape) is testimony that deposition has been an active player in shaping the landscape. Not uncommonly, the width of these valley floors is significant enough to accommodate modest-size towns and substantial agricultural activity.

106 FAST FACT...

DUE TO THE VAST AMOUNT of water as well as sediment that's deposited where the Amazon River meets the Atlantic Ocean, the color and salinity of the Atlantic Ocean are modified for nearly 200 miles (320 kms) from the delta.

107 INTERLOCKING SPURS

AS A RIVER MOVES THROUGH THE UPPER COURSE, it cuts downwards. The gradient here is steep and the river channel is narrow. Vertical erosion in this highland part of the river helps to create steep-sided V-shaped valleys, rapids, waterfalls, gorges and as the river winds and bends to avoid areas of hard rock it creates interlocking spurs, which look a bit like the interlocking parts of a zip.

If the river valley is subsequently subject to glaciation, the glacier shears off the tips of the interlocking spurs, due to its straighter course, creating truncated spurs.

While similar in general appearance, the mechanism behind the formation of interlocking spurs is different to that behind meanders, which arise out of a combination of horizontal erosion and deposition.

108 FAST FACT...

THE YANGTZE RIVER is the longest river in Asia. With its source in Tibet and mouth at Shanghai, it flows for 3,988 miles (6,418 kilometers) and features some spectacular interlocking spurs.

109 OXBOW LAKES

AS RIVERS MEANDER ACROSS WIDE VALLEYS or flat plains, the force of the water starts to erode and undercut the river bank on the outside of the bend where water flow has most energy due to decreased friction. On the inside of the bend, on the other hand, where the river flow is slower, material is deposited, as there is more friction. Over time, this causes the bends to become larger and more circular. Eventually, and especially during floods, the river begins to cut the loop off by eroding the neck of the loop and breaks through to form a whole new streambed.

110 FAST FACT...

IN AUSTRALIA, oxbow lakes are also known as billabongs.

111 FAST FACT...

THE MISSISSIPPI is shorter in length now than in the early 19th century because the U.S. government created their own cut offs and oxbow lakes in order to improve navigation along the river.

Sediment is then deposited on the loop side of the stream, cutting off the loop from the stream entirely. This results in a crescent-shaped lake that looks exactly like an abandoned river meander. Such lakes are called oxbow lakes because they look like the bow part of the yoke used with teams of oxen.

112 WATERFALLS

NEAR THE SOURCE, a river flows over steep slopes with an uneven surface. These highland areas are usually composed of hard igneous rocks, which are ideal for forming waterfalls and rapids, for example.

As water increases its velocity at the edge of a waterfall, it plucks material from the riverbed. Whirlpools created in the turbulence as well as sand and stones carried by the water, increase the erosion capacity, which causes the waterfall to carve deeper into the bed and to recede upstream. Often, over time, the waterfall will recede back to form a canyon. The rate of retreat for a waterfall can be as high as 4.92 ft per year.

Waterfalls can also occur where melt water drops over the edge of a tabular iceberg or ice shelf, or along the edge of a glacial trough, where a stream or river flowing into a glacier continues to flow into a valley after the glacier has receded or melted. This is known as a hanging valley.

Waterfalls are grouped into 10 broad classes based on average volume of water present on the fall (which depends on the waterfalls average flow and height).

114 FAST FACT...

WITH A VERTICAL DROP of more than 165 ft (50 m), Niagara Falls (a Class 10 waterfall) has the highest flow rate of any waterfall in the world.

113 FAST FACT...

ANGEL FALLS in Venezuela is the world's tallest waterfall at 3,212 ft (979 m).

115 FLOODING – THE CAUSES

A FLOOD OCCURS WHEN A RIVER BURSTS ITS BANKS and the water spills onto the floodplain. Flooding tends to be caused by heavy rain; the faster the rainwater reaches the river channel, the more likely it is to flood. The nature of the landscape around a river will influence how quickly rainwater reaches the channel.

The following factors may encourage flooding:

- A steep-sided channel - a river channel surrounded by steep slopes causes fast surface run-off.

- A lack of vegetation or woodland - trees and plants intercept precipitation (i.e., they catch or drink water). If there is little vegetation in the drainage basin then surface run-off will be high.

- A drainage basin, consisting of mainly impermeable rock - this will mean that water can't percolate through the rock layer, and so will run faster over the surface. Urban areas particularly have drainage basins consisting largely of impermeable concrete. In urban areas, drains and sewers also take water quickly and directly to the river channel whilst houses with sloping roofs further increase the amount of run-off.

116 FAST FACT...

THE AMAZON RIVER can be as much as one to six miles wide. This may seem wide but during flood seasons it can be much, much wider; some report it is more than 20 miles wide (32 kms) in certain places.

117 FLOODING – THE IMPACT

FLOODS CAN CAUSE DAMAGE TO HOMES AND POSSESSIONS as well as disruption to communications. However, flooding can also have a positive impact on an area as it deposits fine silt (alluvium) onto the floodplain, making it very fertile and excellent for agriculture. In less developed countries, people living on or near floodplains may rely upon regular flooding to help support their farming and therefore provision of food.

On the other hand, those in less developed countries can be more vulnerable as they don't, if need be, often have the resources to prevent flooding or deal with the aftermath of it.

118 FAST FACT...

MOST OF THE LAND in Bangladesh forms a delta from three main rivers - Ganges, Brahmaputra, and Meghna - and flooding is an annual event. It's beneficial though as it provides water for the rice and jute (two main crops in the area) and also helps to keep the soil fertile.

The Mozambique floods of 2000 show that what happens in one country can very often affect another. South Africa had five weeks of heavy rain early in the year. Botswana was also particularly badly hit, with 75% of its yearly rainfall in three days, which meant the rivers, including the Zambezi eventually burst their banks causing severe flooding before they reached the sea in Mozambique. The results were disastrous; services were cut off and many people were stranded, homeless or had died through drowning or disease.

119 FLOODPLAIN

A FLOODPLAIN REFERS to the flat land beside rivers that is prone to flooding. The extent of floodplains may vary from a few feet to many miles on either side of the water course. The latter is common in the case of large rivers flowing through coastal plains.

A floodplain is a very fertile area due to the rich soil, known as alluvium, deposited by past floodwaters. This makes floodplains a good place for agriculture and has meant that many floodplain regions, including the lower Nile, Ganges, Huang, and Yangtze (Chang) Rivers have long supported very high human-population densities. Whilst floodplains represent a promise land for many, living on a floodplain can have serious pitfalls. Whilst a floodplain can attract and nourish so many, it can also be the scene of terrible flooding with devastating consequences.

120 RIVER MANAGEMENT – HARD ENGINEERING

🎓 **RIVER MANAGEMENT TECHNIQUES** are mainly centered around flood prevention. Commonly, the aim is to lengthen the amount of time it takes for water to reach the river channel, thereby increasing the lag time.

Hard engineering options including dam construction and river engineering tend to be more expensive and have a greater impact on the river and the surrounding landscape.

DAM CONSTRUCTION
- Dams are often built along the course of a river in order to control the amount of discharge.
- Water is usually stored in a reservoir behind the dam. This water can then be used to generate hydroelectric power or for recreation purposes.

RIVER ENGINEERING
- A river channel may be widened or deepened allowing it to carry more water, straightened so that water can travel faster along the course, or altered to divert floodwaters away from settlements. Sometimes though, altering the river channel may lead to a greater risk of flooding downstream, as the water is carried there faster.

121 FAST FACT...

📖 **HIGHEST OR MOST**
expensive dam in the World?
Aswan Dam.

122 RIVER MANAGEMENT – SOFT ENGINEERING

🎓 **SOFT ENGINEERING OPTIONS** are generally more ecologically sensitive and include afforestation and what is known as, managed flooding.

AFFORESTATION
- Trees are planted near to the river. This means greater interception of rainwater and lower river discharge. This is a relatively low cost option, which enhances the environmental quality of the drainage basin.

MANAGED FLOODING
- The river is allowed to flood naturally in places, to prevent flooding in other areas including settlements for example.

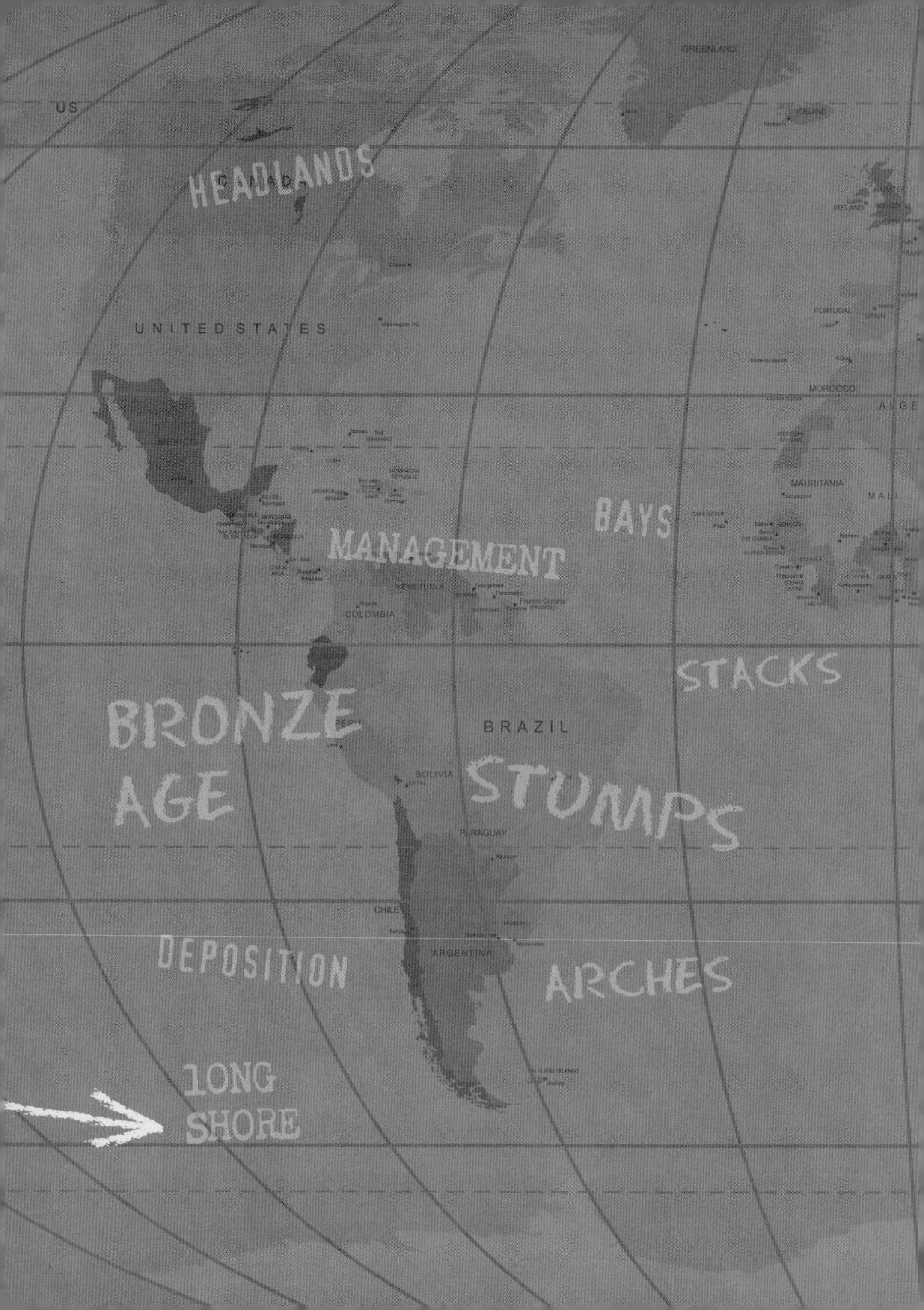
HEADLANDS
BAYS
MANAGEMENT
STACKS
BRONZE AGE
STUMPS
DEPOSITION
ARCHES
LONG SHORE
US
GREENLAND
ICELAND
CANADA
IRELAND
UNITED STATES
PORTUGAL
SPAIN
MOROCCO
ALGE
MEXICO
CUBA
MAURITANIA
MALI
VENEZUELA
COLOMBIA
French Guiana
BRAZIL
BOLIVIA
PARAGUAY
CHILE
ARGENTINA

STACKS

RUSSIA

DRIFT

Coastal

EROSION

CAVES

CLIFFS

WAVE

CAPES

BEACH

123 WAVE ACTION – DESTRUCTIVE OR CONSTRUCTIVE

COASTS ARE SHAPED BY THE SEA and the action of waves is one of the most significant forces of coastal change. Waves are created by wind blowing over the surface of the sea. As the wind blows over the sea, friction is created producing a swell in the water. The energy of the wind causes water particles to rotate inside the swell and this moves the wave forward.

The size and energy of a wave is influenced by how long the wind has been blowing, the strength of the wind and how far the wind has travelled (the fetch).

Waves can be constructive or destructive. When a wave breaks, water is washed up the beach, this is called the swash. Then the water runs back down the beach, which is called the backwash. With a constructive wave, the swash is stronger than the backwash. With a destructive wave, the backwash is stronger than the swash.

Destructive waves, which have a short wave length and are high and steep are created in storm conditions and tend to erode the coast. Constructive waves though, have a longer wavelength and are low in height, and are created in calm weather. They break on the shore and deposit material, building up beaches.

124 FAST FACT...

AT HIGH TIDE, the water will be deepest offshore and larger waves with more energy can reach the beach or cliff. Storm waves at high tide have the highest energy of all.

125 LONGSHORE DRIFT

LONGSHORE DRIFT IS ALL TO DO WITH TRANSPORTATION of materials by wave action. Depending on the direction of the prevailing wind, waves, containing all the various sources of material, can approach the coast at an angle. The swash of the waves then carries the material up the beach at an angle whilst the backwash flows back to the sea in a straight line at 90°. This continual action of swash traveling up the beach at an angle, and backwash flowing back down the beach in a straight line, means that materials are transported sideways along the coast. This movement of material, mainly sediment, is known as longshore drift and occurs in a zigzag along the coast.

There are four ways, namely solution, suspension, saltation and traction, in which waves transport particles from minerals to silts and clays and shingle to large pebbles, all of which can contribute to the movement of sediment by longshore drift.

126 COASTAL EROSION

EVERY WAVE THAT STRIKES LAND potentially performs mechanical weathering and every drop of water within that wave may contribute to chemical weathering. Both are forms of erosion, which may include the wearing away and breaking up of rock along the coast.

Destructive waves erode the coastline in a number of ways:

Hydraulic action – Air may become trapped in joints and cracks on a cliff face. When a wave breaks, the trapped air is compressed which weakens the cliff and causes erosion.

Abrasion – Bits of rock and sand in waves grind down cliff surfaces like sandpaper.

Attrition – Waves smash rocks and pebbles on the shore into each other, and they break and become smoother.

Solution – Acids contained in sea water will dissolve some types of rock such as chalk or limestone.

127 FAST FACT...

SANDY BEACHES ARE MOST AT RISK from severe storms as the tiny particle size of each grain of sand means that waves can easily transport them and greatly diminish a beach.

128 COASTAL DEPOSITION

WHEN A WAVE LOSES ENERGY, IT DROPS THE SAND, rock particles and pebbles it has been carrying. This is called deposition. Deposition happens when the swash is stronger than the backwash and is associated with constructive waves.

Deposition is likely to occur when waves enter an area of shallow water, or a sheltered area such as a cove or bay, when there is little wind and a good supply of materials. Deposition can of course, create an accumulation of sand in the near-shore environment, forming shallow rises called sand bars. Other materials may create or add to spits (small points of land that extend outwards into the water), or long narrow barrier islands that roughly parallel the coast. Barrier islands however represent a huge collection of small particles that can be eroded far more readily than they are deposited.

129 FAST FACT...

WELL-KNOWN BARRIER ISLANDS include New York's Fire Island, South Carolina's Hilton Head and Florida's Cape Canaveral and Miami Beach.

HOW ARE CLIFFS FORMED?

CLIFFS ARE ONE OF THE MOST COMMON FEATURES of a coastline and are shaped through a combination of erosion and weathering (the breakdown of rocks caused by weather conditions).

The rock type greatly affects the cliff formation. Soft rock, which is sand or clay based, erodes easily to create gently sloping cliffs. Hard rock, like chalk, is more resistant though and erodes slowly to create steep cliffs.

Firstly, the weather weakens the top of a cliff then the sea attacks the base forming what is known as a wave-cut notch. Over time, the notch increases in size causing the cliff to collapse. The backwash of the waves then carries the rubble towards the sea forming a wave cut platform. This process repeats time and time again and the cliff continues to retreat.

Wave-cut platforms are often most obvious at low tide when they become visible as huge areas of flat rock.

131 FAST FACT...

 WHILST OFTEN PROVIDING SPECTACULAR VIEWS, a cliff is an erosional feature thus the associated insurance premiums for any cliff front properties.

HEADLANDS AND BAYS

A HEADLAND IS A POINT OF LAND OR PROMONTORY, usually high and often with a sheer drop that extends out into a body of water. Headlands are formed when the sea attacks a section of coast with alternating bands of hard and soft rock.

The bands of soft rock, such as sand and clay, erode more quickly than those of more resistant rock, such as chalk. This leaves a section of land jutting out into the sea called a headland. Bays are formed next to the headland where the soft rock erodes away.

133 FAST FACT...

LARGE HEADLANDS are known as Capes.

134 FAST FACT...

LARGER HEADLANDS will almost certainly have a lighthouse at their point.

Coastlines, namely discordant coastlines alternate between hard and soft rock strata (or bands) hence bays often occur in areas of softer rock as opposed to harder varieties. A concordant coastline on the other hand, has the same type of rock along its length and tends to have fewer bays and headlands.

135 FAST FACT...

FROM CAPE REINGA IN NEW ZEALAND, you can see where the Tasman Sea meets the Pacific Ocean and trace the line where they clash in a confluence of waves and white water.

136 CAVES, ARCHES, STACKS AND STUMPS

THESE FEATURES CAN ALL BE CREATED BY WEATHERING and erosion and occur through a step-by-step process.

Firstly, hydraulic action is the predominant process that creates caves. Waves force their way into cracks in the cliff face causing the sand and other materials to grind away at the rock until the cracks deepen to become caves. If a cave is formed in a headland, it may eventually break through to the other side forming an arch.

Gradually, and with more wave action, that arch becomes bigger until it can no longer support the top of the arch and then collapses. This leaves the headland on one side and a stack (a tall column of rock) on the other. A stack will be attacked, by oncoming waves, at the base, which weakens the structure until it eventually collapses to form a stump.

138 FAST FACT...

STANDING AT 164 FEET HIGH the Azure Window in Gozo, Malta, is an arch giving one of the most spectacular window-like views of the Mediterranean Sea.

137 FAST FACT...

PERHAPS the best known stacks are The Twelve Apostles on the Great Ocean Road in Australia. Due to on-going erosion only 8 of the 12 apostles are left today but it's expected that the existing limestone headland will result in new stacks in the future.

139 COASTAL FLOODING

👨‍🎓 **COASTAL REGIONS ARE PARTICULARLY** vulnerable to flooding and the impacts on coastal communities can be devastating. But, what makes some coastal areas especially vulnerable?

Well, there are several factors that increase the chances of flooding including low lying areas where much of the land is below sea level, where cliffs are formed of soft, easily eroded rocks such as glacial till (boulder clay), where storm surges are more likely, and where there is a lack of adequate hard or soft engineered sea defences of course.

Also, the coastal plain is where rivers reach their peak volume by virtue of so many tributaries adding to the combined flow. Velocity is fairly slow, however, since by definition coastal plains are low-lying and flat. Here individual rivers flow through broad flood plains bound by diminutive valley walls. The rivers themselves are bound by even less diminutive natural levees allowing the water to easily overtop the river banks.

141 FAST FACT...

📖 **A 'FLOOD ACTION PLAN' IS FUNDED BY THE WORLD BANK** to monitor flood levels, and construct flood banks/artificial levees in Bangladesh where coastal flooding is an annual event.

140 FAST FACT...

📖 **1 MILLION PEOPLE** were made homeless and about 1,200 people drowned in the New Orleans flood in 2005.

142 FAST FACT...

📖 **GLOBAL WARMING** will likely increase the chances of storm surges thus increasing the likelihood of coastal flooding.

COASTAL MANAGEMENT — HARD ENGINEERING

🎓 **TO CONTROL NATURAL PROCESSES** such as erosion and longshore drift on coastlines, physical management of the coast is attempted. Just how successful it is depends on understanding the different uses of coastal land and the physical processes impacting on the coast.

Each type of coastal management has its advantages but also its disadvantages. Hard engineering options include a sea wall for example, which protects the base of a cliff, land, and buildings against erosion and can also prevent coastal flooding in some areas. It's expensive to build and to maintain though.

Wooden barriers built at right angles to the beach, known as groynes, can act as defense by preventing the movement of beach material along the coast by longshore drift. This allows a beach to build up, which in itself is a natural defense against erosion. Again, they're costly to build though.

Rock armour or boulder barriers are another option. They're costly to obtain and transport though.

144 FAST FACT...

📖 **COASTAL DEFENSE** is especially important in the UK for example, where over 17 million people live within 10km of the coast.

145 FAST FACT...

📖 **THERE ARE OFTEN CONFLICTING DEMANDS** on our coastlines, including economic and conservation issues.

146 COASTAL MANAGEMENT — SOFT ENGINEERING

SOFT ENGINEERING options are often less expensive than hard engineering. They're also usually more long-term and sustainable, with less impact on the environment although they may require more maintenance.

Beach management

- replaces beach or cliff material that has been removed by erosion or longshore drift.

- beaches act as a natural defense against erosion and coastal flooding.

- a relatively inexpensive option but requires constant maintenance to replace the beach material as it's washed away.

Managed retreat

- areas of the coast are allowed to erode and flood naturally. Usually this will be areas considered to be of low value – i.e., places not being used for housing or farmland.

- advantages are that it encourages the development of beaches (a natural defense) and salt marshes (important for the environment) and cost is low.

- a cheap option, although there may be some compensation needed for loss of buildings and farmland.

147 FAST FACT...

THE INCREASED THREAT of a rise in sea levels due to climate change will mean that many more areas will need to consider the sustainability of coastal defenses for the future.

ATLANTIC
MUMMIES
DEEP
MEDITERRANEAN
GULF STREAM
EL NIÑO
OCEANOGRAPHY
CONTINENTAL SHELF
GREENLAND
ICELAND
CANADA
UNITED STATES
IRELAND
PORTUGAL
SPAIN
MOROCCO
MAURITANIA
MALI
MEXICO
CUBA
VENEZUELA
COLOMBIA
French Guiana
PERU
BRAZIL
BOLIVIA
PARAGUAY
CHILE
ARGENTINA

Seas & Oceans

148 WATERY WORLD

✦ MORE THAN 70% OF THE EARTH

is covered by water and the water, or hydrological cycle, is the continuous movement of this water between the sea, the atmosphere and the land.

The water cycle is known as a closed system, i.e., no amount of water is lost or gained instead it's recycled again and again through the process of evaporation, condensation, precipitation, and water transfers such as surface run off. In a way it's the surface run-off that completes the water cycle. Some of the water that falls to Earth collects on the surface and begins a down-slope journey to the sea. This is essentially fresh water on the move and serves as a water supply system for the population.

149 FAST FACT...

📖 THE SUN IS VITAL to the

water cycle; acting like a pump to set the cycle in motion.

Vegetation affects the water cycle, as for example, in the summer months when woodlands and forests become increasingly lush, rainfall is intercepted. Although rain isn't reaching the soil, the cycle isn't stopped it just changes slightly. Holding moisture, plants are caused to sweat by the hot sun, therefore providing key input into the atmospheric vapor, especially in tropical areas.

150 FAST FACT...

📖 SEAS AND OCEANS contain

97% of the world's water, and ice holds 2%. That leaves just 1% of the world's water as fresh water on land or in the air.

THE PACIFIC OCEAN IS BY FAR THE WORLD'S largest ocean at 60,060, 700 square miles (155,557,000 sq kms). It covers approximately 28% of the Earth and is equal in size to nearly all of the land area on the Earth. Located between the Southern Ocean, Asia and Australia and the Western Hemisphere, the Pacific Ocean has an average depth of 13,215 feet.

The Atlantic Ocean is the next biggest with an area of 29,637,900 square miles (76,762,000 sq kms). It's located between Africa, Europe, the Southern Ocean and the Western Hemisphere.

The Indian Ocean is the world's third-largest ocean and it has an area of 26,469,900 square miles (68,566,000 sq kms). It is located between Africa, the Southern Ocean, Asia, and Australia.

In 2000, the International Hydrographic Organization defined the world's newest and fourth-largest ocean, the Southern Ocean. In doing so, boundaries were taken from the Pacific, Atlantic, and Indian Oceans. It has a total area of 7,848,300 square miles (20,327,000 sq kms).

The Arctic Ocean with an area of 5,427,000 square miles (14,056,000 sq kms) extends between Europe, Asia, and North America (most of its waters are north of the Arctic Circle) is the world's smallest ocean.

152 FAST FACT...

THROUGHOUT most of the year, much of the Arctic Ocean is covered by a drifting polar icepack that's an average of 10 feet thick.

153 FAST FACT...

OCEANOGRAPHIC RESEARCH identified the 'Southern Circulation' as one of the main drivers of ocean systems, which sets the body of water now known as the Southern Ocean apart as a separate ecosystem.

154 WHAT FEEDS OUR OCEANS?

ANY NUMBER OF RIVERS FLOW INTO THE PACIFIC OCEAN INCLUDING
The Yangtze, Fraser, Columbia, Yukon, etc., and similarly lots empty in the
Atlantic Ocean including those on the eastern coast of South America, many on
the eastern coast of North America, and those on the western coast of parts of
Europe as well as the western coast of Africa. The Amazon River is surely the
best-known river emptying, in South America, into the Atlantic Ocean.

Because of its location, any river flowing out of the east coasts of South Africa,
Mozambique, Tanzania, Kenya, Somalia, Yemen, or Oman; any river flowing
out of Madagascar; any river flowing out of the south coasts of India, Pakistan,
Bangladesh, Burma or any river flowing out of the west coast of Indonesia ends
up in the Indian Ocean. Also the Persian Gulf, Gulf of Oman, Red Sea, and Gulf of
Aden go to the Indian Ocean.

155 FAST FACT...

THE AMAZON RIVER is
responsible for about 20% of the
fresh water that flows into the
world's oceans.

OCEANS CURRENTS — SURFACE

EACH OF THE FIVE BIGGEST OCEANS ARE CONNECTED DUE TO OCEAN CURRENTS, which circulate water all over the globe. These currents play a major role when it comes to climate. For example, warm surface currents act like a global heating system bringing significant warmth to high latitude areas that would otherwise be much cooler whilst cold surface currents act like air-conditioning units causing low latitude areas to be much cooler.

To a large extent, ocean currents also determine the global geography of precipitation. The Sun evaporates warm water more easily than cold and therefore produces the atmospheric vapor that results in rain more readily in warmer climates.

157 FAST FACT...

THE GULF STREAM is a warm-water current that moves up the Eastern coast of the United States and then becomes the North Atlantic Current, which reaches Europe with a considerable amount of stored heat remaining.

Generally, surface currents exhibit circular movements. North of the Equator, the flow is usually clockwise and south of the Equator, the flow tends to be counter-clockwise. As ocean currents move westward along the Equator, they absorb lots of solar energy, heat up, and become warm currents. As they turn away from the Equator, they generally continue to absorb about as much heat as they dissipate, at least while they remain in the Tropics.

After leaving the Tropics, the reverse starts to happen, i.e., the currents slowly radiate more heat than they gain, which means that they remain comparatively warm for longer.

158 FAST FACT...

SURFACE CURRENTS are mostly caused by wind because it creates friction as it moves over the water.

159 OCEAN CURRENTS — DEEP

DEEP WATER CURRENTS, ALSO CALLED THERMOHALINE CIRCULATION, are found below 1300 feet and make up about 90% of the ocean. Like surface currents, gravity plays a role in the creation of deep water currents although deep water currents are mainly caused by density differences in the water.

Density differences are a function of temperature and salinity. Warm water for example, holds less salt than cold water so is less dense and rises toward the surface while cold, salt-laden water sinks. As the warm water rises, the cold water is forced to rise through something called 'upwelling' and fill the void left by the warm. When cold water rises though, it too leaves a void and the rising warm water is then forced, through 'downwelling', to descend and fill this empty space. It's this process and the associated movement that causes deep water currents.

160 FAST FACT...

DEEP WATER CURRENTS are known as the 'Global Conveyor Belt' because they are what moves water throughout the oceans.

161 FAST FACT...

THE WORLD'S LARGEST OCEAN CURRENT is the Antarctic Circumpolar Current that moves east in the Southern Ocean.

162 EL NINO & LA NINA

IN ANY GIVEN YEAR, THERE ARE ANOMALIES when it comes to climate and average yearly conditions. El Niño and La Niña (so-called because Niño means boy in Spanish and Niña means girl), which happen every so many years, are good examples of this.

Whilst surface currents are usually cool in the tropical portion of the Pacific, during an El Niño, they become unusually warm. The affected ocean water circulates and also influences the behavior of atmospheric pressure belts, the impact of which can be substantial and widespread. Just what that means varies from place to place and year to year. Sometimes, for example, rainy seasons become extremely stormy and dry seasons become prolonged droughts. The effects aren't always bad though, as may be evidenced by a normally harsh winter that turns mild.

The reasons for El Niño and La Niña aren't fully understood but because the conditions occur around Christmas in the waters off western South America, the local populace named El Niño, referring to the Christ child. During La Niña, the opposite happens (girl being the opposite of boy) and the water becomes unusually cold.

164 FAST FACT...

📖 **VERY STRONG EL NIÑO** events in 1965-1966, 1982-1983, and 1997-1998 caused significant flooding and damage from California to Mexico to Chile.

163 FAST FACT...

📖 **EL NIÑO OCCURS** every two to five years, and can last for several months or even a few years.

165 TSUNAMI

🎓 **A TSUNAMIS ISN'T A TIDAL WAVE AS MANY IMAGINE,** but instead it's a series of powerful ocean waves. The most common cause of tsunamis are earthquakes although they can also be caused by volcanic eruptions, landslides, and underwater explosions.

When an earthquake occurs below the ocean's surface, with a magnitude large enough to cause disturbance on the ocean floor and move a significant amount of material, the water surrounding the disturbance is displaced and radiates away from the initial source (i.e., the epicenter in an earthquake) in a series of fast moving waves. In addition, in the case of an earthquake, its magnitude, depth, water depth, and the speed at which the material moves all factor into whether or not a tsunami is generated.

Tsunamis' can travel thousands of miles at speeds of up to 500 miles/hour (805 kms/hour). If a tsunami is generated in the deep ocean, the waves radiate out from the source of the disturbance and move toward land on all sides. These waves usually have a large wavelength and a short wave height so they are not easily recognized by the human eye.

As the tsunami moves toward shore and the ocean's depth decreases, its speed slows quickly and the waves begin to grow in height as the wavelength decreases. This is called amplification and it is when the tsunami is the most visible and can have the most devastating effects.

166 FAST FACT...

📖 **TSUNAMIS** is Japanese for 'harbor wave'.

167 FAST FACT...

📖 **IN AN EFFORT** to generate stronger warning systems for Tsunamis, there are monitors throughout the world's oceans to measure wave height and potential underwater disturbances.

168 PACIFIC OCEAN — WORLD'S BIGGEST

🎓 **THE PACIFIC OCEAN IS THE LARGEST OF THE EARTH'S OCEANIC DIVISIONS.** It extends from the Arctic in the north to the Southern Ocean (or, depending on definition, to Antarctica) in the south, bounded by Asia and Australia in the west, and the Americas in the east.

The Equator subdivides it into the North Pacific Ocean and South Pacific Ocean, with two exceptions: the Galápagos and Gilbert Islands, while straddling the Equator, are deemed wholly within the South Pacific.

170 FAST FACT...

📖 **THE PACIFIC OCEAN** is surrounded by a zone of violent volcanic and earthquake activity sometimes referred to as the Pacific Ring of Fire.

At 165.25 million square kilometers (63.8 million square miles) in area, the Pacific covers about 46% of the Earth's water surface and about one-third of its total surface area, making it larger than all of the Earth's land area combined.

169 FAST FACT...

📖 **THERE ARE SOMEWHERE** between 20,000 and 30,000 islands in the Pacific Ocean. Some of them are high islands, with volcanic, fertile soil. Others are low islands or reefs with less fertile soil.

The Pacific Ocean got its name by Portuguese explorer Ferdinand Magellan during the Spanish expedition of world circumnavigation in 1521, who, on reaching the ocean encountered favorable winds and so named it Mar Pacifico, which is Portuguese for 'peaceful sea'.

171 THE CHALLENGER DEEP — WORLD'S DEEPEST OCEAN TRENCH

AT NEARLY 7 MILES DEEP, the Mariana Trench in the Pacific Ocean is the world's deepest ocean trench and The Challenger Deep, a small slot-shaped depression in the floor of the trench, is the deepest known point.

But what makes it so deep? Well it's the combination of oceanic plates found at the Mariana Trench that makes it so extreme. Unlike a continental collision where mountain ranges like the Himalayas or the Alps are pushed up, Oceanic plates don't collide, but cause one another to suck downwards.

In this case, the Pacific plate is moving westward and plunging below the Philippine plate. The Philippine plate is young and soft though, and as a result, it gets pulled down, along with the sinking Pacific slab.

This combined with the fact that the trench is far from land means it doesn't fill with sediment like many trenches do, and it becomes extremely deep.

172 FAST FACT...

THE CHALLENGER DEEP is named after the British Royal Navy survey ship, HMS Challenger, whose expedition in the 1870s made the first recordings of its depth.

173 FAST FACT...

ONLY FOUR DESCENTS to The Challenger Deep have ever been made and the most recent was a manned solo descent in 2012.

174 CONTINENTAL SHELF — A THRIVING ENVIRONMENT

🎓 **A CONTINENTAL SHELF IS JUST AS YOU MIGHT EXPECT;** a shelf, or relatively flat expanse of ocean bottom found off the coast of continents. The distance a continental shelf extends from the shore varies (it can be up to 200 miles) but the average depth is about 300 feet, which is of course considered shallow by ocean standards. The waters causing a continental shelf to be submerged are known as shelf seas.

Continental shelves represent a valuable resource when it comes to foodstuff (fish are commonly in abundance over the continental shelf) and minerals for example, and with a growing population and higher demand for food, continental shelves are being looked to as sources of supply.

Not just foodstuff but petroleum and natural gas are also found under the continental shelves hence the prevalence of offshore oil rigs in these areas. In this respect, continental shelves represent important economic activity.

175 FAST FACT...

📖 **DUE TO THE SHALLOW** water on Continental Shelves, sunlight is able to reach the ocean floor and gives rise to plant life that serves as food for fish to thrive.

176 FAST FACT...

📖 **THE SIBERIAN SHELF** in the Arctic Ocean is the biggest continental shelf in the World.

177 FIVE BIGGEST SEAS

A SEA IS DEFINED AS A LARGE lake-type water body that has saltwater and is sometimes, although not always, attached to an ocean.

The Mediterranean Sea is the biggest sea of all with an area of 1,144,800 square miles (2,965,800 sq kms) and an average depth of 4,688 feet (1,429 m). The Mediterranean sea is attached to the Atlantic Ocean.

The Caribbean Sea, also attached to the Atlantic, is the next largest with an area of 1,049,500 square miles (2,718,200 sq kms). Although it's not as big as the Mediterranean, its average depth is greater at 8,685 feet (2,647 m).

178 FAST FACT...

THE CASPIAN SEA is an example of an inland sea that isn't attached to an ocean.

The South China Sea with an area of 895,400 square miles (2,319,000 sq kms) is the third biggest sea. It's attached to the Pacific Ocean as is the Bering Sea, which covers a marginally smaller area of 884,900 square miles (2,291,900 sq kms). Finally, the Gulf of Mexico is the fifth biggest sea, which, attached to the Atlantic is 615,000 square miles (1,592,800 sq kms).

179 FAST FACT...

THE CARIBBEAN SEA is littered with over 7,000 islands, islets (very small rocky islands), coral reefs and cays (small, sandy islands above coral reefs).

180 MEDITERRANEAN SEA

THE MEDITERRANEAN SEA IS A VERY LARGE SEA that is bounded by Europe, Africa and Asia, and stretches from the Strait of Gibraltar on the west to the Dardanelles and the Suez Canal on the east. The Mediterranean is almost completely enclosed aside from these narrow locations. The Strait of Gibraltar for example, between Spain and Morocco, which joins the Mediterranean to the Atlantic Ocean, is just 14 miles (22 kms) wide. Because it's almost landlocked the Mediterranean has very limited tides.

Geographically, the Mediterranean is divided into two different basins and it total it borders 21 different nations as well as several different territories. Some of the nations with borders along the Mediterranean include Spain, France, Monaco, Malta, Turkey, Lebanon, Israel, Egypt, Libya, Tunisia and Morocco. It also borders several smaller seas and is home to over 3,000 islands. The largest of these islands are Sicily, Sardinia, Corsica, Cyprus, and Crete.

181 FAST FACT...

IT'S BELIEVED THAT the Egyptians were the first to begin sailing the Mediterranean Sea in 3000 B.C.E.

182 FAST FACT...

7 MILLION YEARS AGO, the Mediterranean totally dried up turning instead into a vast desert where only a thin crust of salt was left.

183 WHY ARE OUR OCEANS SALTY?

THE SALT IN OUR OCEANS is the result of millions of years of minerals leaching and dissolving from the solid Earth. The major portion comes from rivers. As water flows in rivers, it picks up small amounts of mineral salts from the rocks and soil of river beds. This very-slightly salty water flows into the oceans and seas. Other salt is dissolved from rocks and sediments below the ocean floor, through volcanic vents. The weather is also responsible, although to a lesser degree, as rain also deposits mineral particles into the oceans.

185 FAST FACT...

THE LEAST SALTY SEAS are in polar regions, where both melting polar ice and a lot of rain dilute salinity.

Heat from the Sun then distils or vaporises almost pure water from the surface of the sea, leaving the salts and minerals behind. The remaining water then gets saltier and saltier.

As time goes by, and as part of the Earth's water, or hydrological cycle, the evaporated seawater gets returned to the ocean, via rivers or precipitation, to wash down more salt, which becomes ever more concentrated.

184 FAST FACT...

VERY HIGH EVAPORATION rates and low fresh water influx, means that the Red Sea and the Persian Gulf have the saltiest seawater in the World.

186   RISING SEA LEVELS

🎓 **RISING SEA LEVELS** is nothing new. In fact, for the past 18,000 years or so, the world map has been changing, with the ocean on the up-and-up. This is because 18,000 years ago was the peak of the last Ice-Age when much water was 'locked-up' on land in the great glaciers of the world. The ice caps of Greenland and Antarctica plus the majority of other significant glaciers have been slowly shrinking since

then. As the glaciers slowly recede (a byproduct of global warming), their melt water returns to the oceans, which in turn, rise.

The future of global warming isn't yet fully known so just how high the oceans will rise is a matter of on-going debate. Rise, they will though, although the outcome is unknown. Even if the change is slight, the impact will be significant. Island republics such as The Maldives in the Indian Ocean, or Bangladesh, one of the most densely populated areas in the World, where the highest elevation is a mere couple of feet above sea level, are most at risk.

187   FAST FACT...

📖 **18,000 YEARS AGO**, sea levels were approximately 350 feet lower than they are today.

188 OCEAN OWNERSHIP — A CONTROVERSIAL TOPIC

LARGELY BECAUSE OF A WEALTH OF MARINE LIFE, the control and ownership of the world's oceans has long been a controversial topic. Since ancient empires began to sail and trade over the seas, command of coastal areas has been important to governments. However, it wasn't until the 20th century that countries began to come together to discuss a standardization of maritime boundaries. The process has however been problematic and boundaries are still left undefined in many cases.

One of the principal questions is; how far, if at all, offshore, does a nation's sovereign territory extend? It's important to define boundaries not only for ownership of oceanic resources but also for National Security (countries have a right to defend themselves from attack or intrusion), defining police and coastguard jurisdictions, the application of trade and commerce (captains and navigators of freighters and tankers for example, need to know where and when permission is required by virtue of having entered another country's territorial waters) and the ownership of straits (narrow waterways that separate land bodies and which are often major bottlenecks as far as international shipping is concerned).

FAST FACT...

📖 **IN 1992, A TREATY** entitled, the United Nations Convention on the Law of the Seas (UNCLOS) detailing offshore zones or jurisdictions became international law.

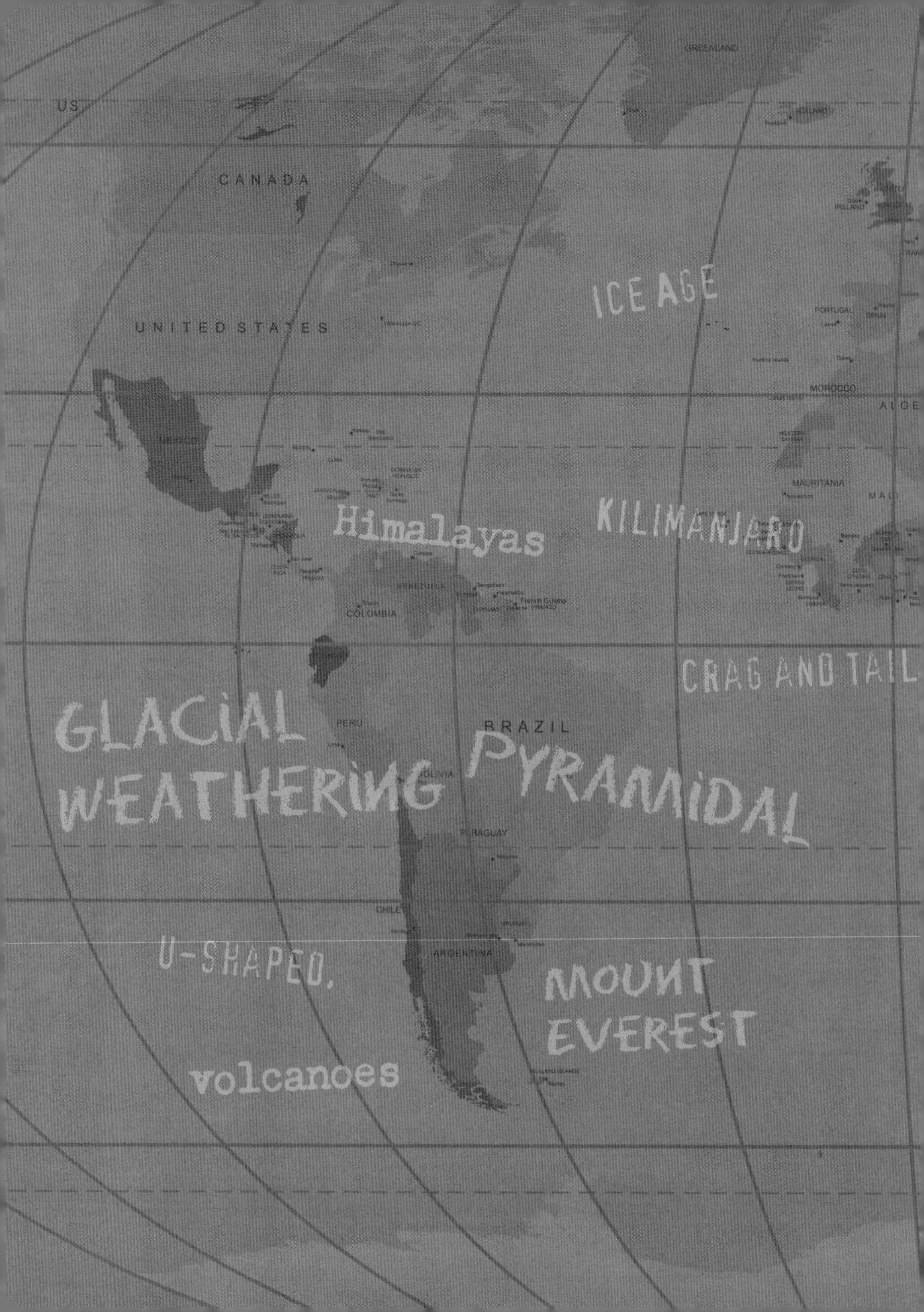
ICE AGE
Himalayas
KILIMANJARO
CRAG AND TAIL
GLACIAL
WEATHERING
PYRAMIDAL
U-SHAPED,
MOUNT
EVEREST
volcanoes
GREENLAND
US
CANADA
UNITED STATES
PORTUGAL
SPAIN
MOROCCO
MAURITANIA
MALI
IRELAND
CUBA
COLOMBIA
VENEZUELA
PERU
BRAZIL
BOLIVIA
PARAGUAY
CHILE
ARGENTINA

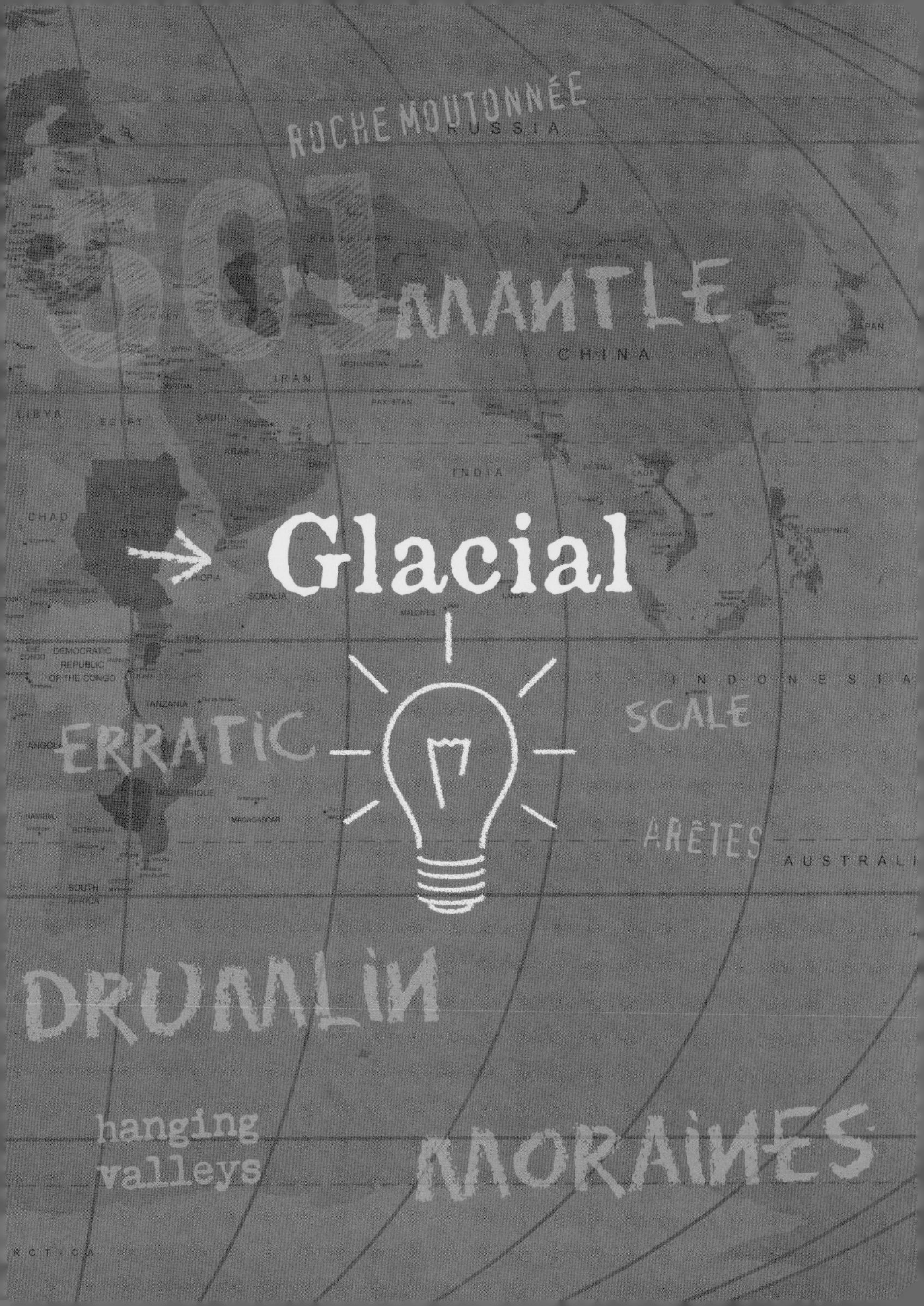

ROCHE MOUTONNÉE

MANTLE

→ Glacial

ERRATIC

SCALE

ARÊTES

DRUMLIN

hanging valleys

MORAINES

190 ICE AGE

ICE AGES were periods in Earth's history when vast glaciers covered large portions of the planet's surface. But, how did this happen? The exact cause of an ice age has not been proven but in very simple terms, whilst the average annual temperature varies constantly from year to year, from decade to decade, and from century to century, for a given period in history, it dropped low enough to allow fields of ice to grow and cover the land.

Over the last 2.5 million years, approximately 24 ice ages have occurred. In each case, an episode of significant warming, called an interglacial period, followed. Earth is currently in an interglacial period. The big question is; are we approaching another Ice Age?

Some scientists believe that an increase in global temperature, as we are now experiencing, could be a sign of an impending ice age and could actually increase the amount of ice on the earth's surface.

Our short history on earth and our shorter record of the climate keeps us from fully understanding the implications of global warming. Without a doubt though, an increase in the earth's temperature will have major consequences for all life on Earth.

191 FAST FACT...

📖 **GLACIERS FOUND** in equatorial regions like South America's Andes Mountains and Mount Kilimanjaro in Africa are remnants of the last Ice Age.

192 WHAT IS A GLACIER AND HOW DOES IT FORM?

🎓 **A GLACIER** is essentially a huge mass of ice resting on land or floating in the sea next to land. Moving extremely slowly, a glacier acts similarly to an immense river of ice, often merging with other glaciers in a stream-like manner.

Regions with continuous snowfall and constant freezing temperatures foster the development of these frozen rivers. It is so cold in these regions that when a snowflake hits the ground it does not melt, but instead combines with other snowflakes to form larger grains of ice. As more and more snow accumulates, mounting weight and pressure squeeze these grains of ice together to form a glacier.

The formation of glaciers and the process by which they shape the landscape around them is called glaciation. Glaciers are found at high altitude across the globe, even in high mountain regions such as the Himalayas of Southern Asia or the Alps of Western Europe where regular snow and extremely cold temperatures are present, and at lower altitude in high latitudes close to the North and South Poles.

193 MOUNTAIN AND CONTINENTAL GLACIERS

🎓 **AS THE NAMES SUGGEST,** mountain (or alpine) glaciers originate in snow that falls in mountainous areas and continental glaciers build up over large land masses in general.

Mountain glaciers –
When large ice masses form, mountain slopes facilitate their downhill movement, and thus their power to erode and transform the landscape. Further upslope, the erosional power of the ice gnaws away at the mountain, turning rounded tops into pointed horns and basic ridges into jagged crests called arêtes. Glaciated mountains have a spectacularly rugged look about them therefore tending to attract tourists.

194 FAST FACT...

📖 **THE WORLD'S LARGEST** alpine glacier is the Siachen Glacier in the Karakoram Mountains of Pakistan.

Continental glaciers –
There are three primary subtypes of continental glacier including ice sheets, which are the largest of any glacier type. Ice sheets extend over 19,300 square miles and are so heavy that they literally bend the continental crust on which they sit, a phenomena known as isostatic depression. Similar to ice sheets, ice caps are smaller and form a roughly circular, dome-like structure that completely blankets the landscape underneath. Ice fields are those ice caps (although smaller versions) that fail to cover the land and are elongated relative to the underlying topography.

195 FAST FACT...

📖 **THE ICE CAP** of Antarctica which is more than 2 miles thick in some places, is home to 92% of all glacial ice worldwide.

GLACIAL WEATHERING –
BY ABRASION, PLUCKING AND FREEZE-THAW ACTION

🎓 **GLACIAL WEATHERING,** including abrasion, plucking and freeze-thaw action has significant effect on the landscape.

The predominant process is freeze-thaw weathering, which describes the action of glacial melt water on joints, cracks, and hollows in rock. When the temperature reaches freezing point, the water inside cracks freezes, expands, and causes the cracks to widen. When the temperature rises though, the water thaws and contracts, eventually causing rocks to break up and form angular rock fragments.

Plucking occurs when rocks and stones become frozen to the base or sides of the glacier and are plucked from the ground or rock face as the glacier moves, leaving behind a jagged landscape.

Abrasion occurs when rocks and stones become embedded in the base and sides of the glacier. These are then rubbed against the bedrock and rock faces at the sides of the glacier, acting a bit like sandpaper, as the glacier moves and thus causing the wearing away of the landscape. Abrasion leaves behind smooth polished surfaces.

197 **FAST FACT...**

📖 **FOR FREEZE-THAW WEATHERING** to take effect the air temperature needs to fluctuate around freezing point.

198 GLACIAL EROSION – CORRIES

🎓 **GLACIERS EXERT COLOSSAL FORCES** on the land and, through erosion, are responsible for the creation of interesting features, many of which are inter-linked.

Corries, also known as cwms or cirques are deep, semi-circular hollows with very steep, precipitous sides. They're eroded from small hollows high up on the mountainside. Snow collects in the hollow and becomes compressed until the air is squeezed out to become firn (a dense type of snow which has thawed and then compacted and recrystallized) or neve (a less dense version of firn), which eventually turns into glacier ice.

Erosion and weathering gradually makes the hollow bigger and the ice begins to move in a circular motion before it pulls away from the back wall of the hollow creating what is known as a crevasse. Plucked debris from the back wall causes further erosion, which further deepens the corrie.

199 FAST FACT...

📖 **A CREVASSE** is a deep crack in a glacier or ice sheet.

200 FAST FACT...

📖 **WHEN ICE IN A CORRIE** melts, a circular lake, known as a tarn, is often formed at the bottom of the hollow.

201 GLACIAL EROSION – ARÊTES AND PYRAMIDAL PEAKS

🎓 **ARÊTES AND PYRAMIDAL PEAKS** are glacial features formed through erosion and are inter-linked with corries.

An arête is a knife-edge ridge that's formed when two neighboring corries run back to back. As each glacier erodes either side of the ridge, the edge becomes steeper and the ridge becomes narrower.

A pyramidal peak is, in a way, an extension of this as it's formed where three or more corries and arêtes meet. The glaciers have carved away at the top of a mountain, creating a sharply pointed summit.

202 FAST FACT...

📖 **MONT BLANC,** The Matterhorn and Mount Everest are all great examples of pyramidal peaks.

203 GLACIAL EROSION – U-SHAPED AND HANGING VALLEYS

A GLACIAL TROUGH OR U-SHAPED VALLEY is formed when a glacier deepens, widens and straightens a former V-shaped river valley. As opposed to the deep-sided walls of a V-shaped valley, the erosive action of the glacier creates a U-shape with a flat floor and steep sides.

Just like rivers, glaciers have tributaries (small rivers that join the main river channel) and as the main glacier erodes deeper into the valley, tributaries are left higher up the steep sides of the glacier. Having not deepened as much as the main glacial U-shaped valley, the tributaries are left carving out what are known as, hanging valleys, high up on the valley sides. These hanging valleys create waterfalls or steep gorges at the cliff-face.

204 FAST FACT...

A GLACIER cuts through the ridges of interlocking spurs, created as rivers erode the landscape, leaving behind truncated spurs to form deep valley sides.

205 GLACIAL EROSION – VALLEY FLOOR LANDFORMS

🎓 **AS A GLACIER FLOWS** over the land, it travels over hard and soft rock. Softer rock is less resistant, so a glacier carves a deeper trough. When the glacier has retreated, water collects in the deeper area and creates a long, thin lake called a ribbon lake.

The areas of harder rock that are left behind are called rock steps. If a glacier hits a particularly resistant outcrop of rock, it'll flow over and around it. This leaves a rock mount smoothed by abrasion from the glacier. These come in two types:

- **ROCHES MOUTONNÉE** often have steep, jagged faces created by glacial plucking and a gradual incline which is smoothed and polished by abrasion. A Roches moutonnée may have striations on it indicating the direction of glacier movement.

- **CRAG AND TAIL DEVELOPS** when glacial ice hits the steep jagged face of hard rock first therefore protecting one side of the hard rock outcrop. For this reason, a crag and tail tends to be larger than a roche moutonnée.

206 FAST FACT...

📖 **IN CENTRAL PARK,** New York, there are many examples of roche moutonnée where the rock face is marked with striations.

207 FAST FACT...

📖 **EDINBURGH CASTLE** in Scotland is built on a crag and tail.

208 GLACIAL DEPOSITION – ERRATICS AND DRUMLINS

👆 **THE LARGE AMOUNT** of rock carried by a glacier is known as its load and includes everything from large boulders to angular fragments of rock and fine clay.

As a glacier retreats, it deposits its load, called glacial till, creating different landforms. Unlike river deposits that are often sorted into different sizes, all glacial deposits are angular and mixed up, or unsorted.

Erratic - an erratic, which is a large rock or boulder often found on its own, rather than in piles. An erratic is unusually large as well as being an unusual shape and is of a rock type uncommon to the area that it's been dumped in.

Drumlin – a drumlin is an elongated hill of glacial deposits, which can be up to half a mile long and 1640.42 ft wide. The long axis of the drumlin indicates the direction in which the glacier was moving. Glaciologists still disagree as to exactly how Drumlins formed but we do know that they often occur in groups and would have been deposited when a glacier became overloaded with sediment.

209 FAST FACT...

📖 **AN ERRATIC** can be found far away from its source. For example, erratics originating in Norway are found in South East England.

210 FAST FACT...

📖 **THE TOWN OF OKOTOKS** (meaning 'big rock' in the local Blackfoot language) in Alberta, Canada is named after the erratic found nearby. "Big Rock," a brand of beer is even named after it!

GLACIAL DEPOSITION – MORAINES

WHEN GLACIAL ICE MELTS PILES OF ROCKS are laid down that have been carried along by the glacier. These piles are called moraines and there are various types:

- **TERMINAL MORAINES** are found at the terminus or the furthest (end) point reached by a glacier.
- **LATERAL MORAINES** are found deposited along the sides of the glacier.
- **MEDIAL MORAINES** are found at the junction between two glaciers.
- **GROUND MORAINES** are disorganized piles of rocks of various shapes, sizes, and of differing rock types.
- **RECESSIONAL MORAINES** are located behind the outermost edge of a glacier and formed when the glacier lingers in one spot for a long time.
- **PUSH MORAINES** are created by glacial till that was a moraine deposited by an earlier glacier that once covered the area.
- **ABLATION MORAINES** are formed from material that fell upon the glacier.

212 FAST FACT...

MORAINE is a French word that refers to any glacier-formed accumulation.

213 AVALANCHES – CAUSE AND EFFECT

AN AVALANCHE is a sudden downhill movement of snow and is a significant hazard to people living in, or visiting, glacial areas. A slab avalanche is the most dangerous form of movement.

So, what causes an avalanche? Essentially, an avalanche occurs when the gravity pushing the collection of snow at the top of a slope is greater than the strength of the snow itself. A number of factors might be involved in this including heavy snowfall, deforestation (which makes a slope less stable), the presence of steep slopes (helps to increase the speed of movement), vibrations (e.g., from an earthquake, noise or even off-piste skiers), layering of snow (e.g., where snow is already on the mountain and has turned into ice, and then fresh snow falls on top, which can easily slide down) and the direction of wind, which may cause snow to pile up and overhang a mountain.

As for the effects of an avalanche; they can cause widespread disruption, damage and sometimes loss of life. An avalanche is able to obstruct anything in its path. Roads and railways may be blocked for example, and power supplies cut off. A powerful avalanche can even destroy buildings and people can also be killed.

214 FAST FACT...

90% OF ALL avalanches occur on moderate slopes with an angle of 30° to 45° (snow tends not to accumulate on steeper slopes).

215 FAST FACT...

WITH THE GROWTH of winter-time recreation sports, fatalities caused by avalanches have been on the rise since the 1950s.

216 FAST FACT...

📖 **INTERNATIONALLY,** the Alpine countries of France, Austria, Switzerland, and Italy experience the greatest number of avalanches and loss of life annually.

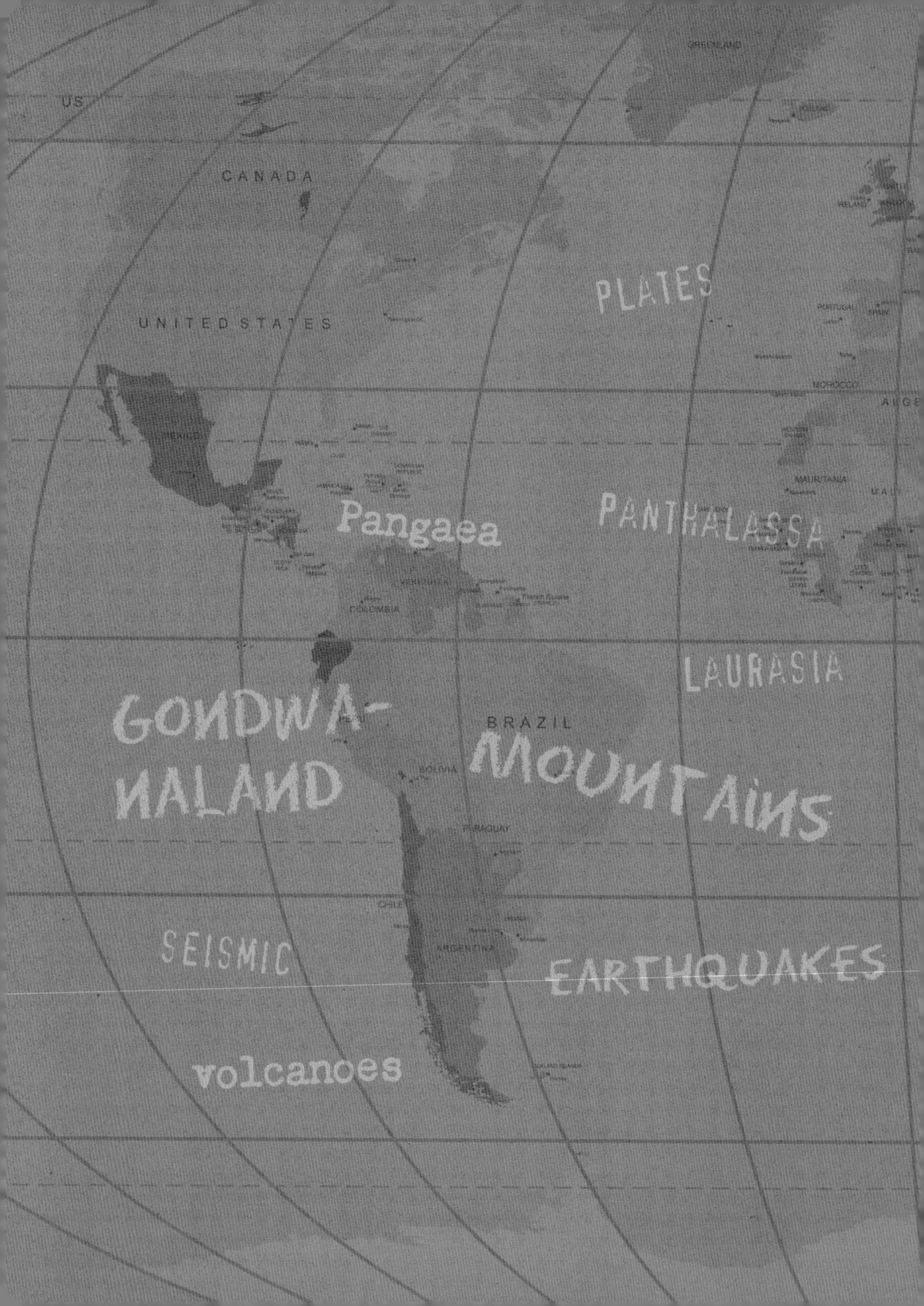

GREENLAND
US
CANADA
IRELAND
PLATES
UNITED STATES
PORTUGAL
SPAIN
MOROCCO
MAURITANIA
Pangaea
PANTHALASSA
VENEZUELA
COLOMBIA
LAURASIA
GONDWA-
NALAND
BRAZIL
MOUNTAINS
BOLIVIA
PARAGUAY
CHILE
ARGENTINA
SEISMIC
EARTHQUAKES
volcanoes

Tectonics

217 PLATE TECTONICS

PLATE TECTONICS IS THE THEORY that explains the movement, formation and destruction of the plates that make up the Earth's crust.

The place where two plates meet is called a plate margin and as plates move, stress and friction occurs along the plate margins thus creating what becomes known as Constructive and Destructive Plate Boundaries.

At constructive plate boundaries, the plates move apart and at destructive plate boundaries the plates move towards each other (this usually involves a continental plate and an oceanic plate). But, what causes the plates to move? Well, radioactive decay in the core of the Earth causes hot circular convention currents to rise and cold currents to fall. It's these currents that cause the motion of tectonic plates. Where convection currents diverge near the Earth's crust, plates move apart and where convection currents converge, plates move towards each other.

218 FAST FACT...

THE CRUST FORMS the outer surface of the Earth and is sub-divided into Oceanic and Continental plates.

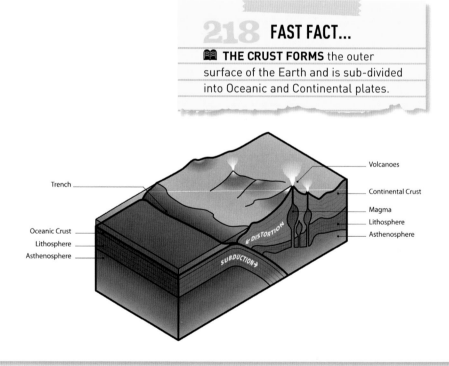

WAY BACK IN 1915, German meteorologist Alfred Wegener popularized a bold theory in which the continents are moving around the Earth's surface, much like ice cubes in a punchbowl. Wegener named his theory 'continental drift'.

According to Wegener, Earth's surface once consisted of a single supercontinent called Pangaea global (meaning 'all the Earth') and a single world ocean, called Panthalassa ('all the seas') until Pangaea eventually broke into two pieces of roughly equal size; a northern component called Laurasia and a southern component called Gondwanaland. These pieces, Wegener suggested subsequently drifted apart, hence the name continental drift.

Over the years, Wegeners 'continental drift' theory came up against much criticism and remained unproven until after his death. Today though, you only need to look at a world map to see that the Atlantic coastlines of Africa and Europe on the one hand and North and South America on the other hand are like giant pieces of a jigsaw, separated by the Atlantic Ocean.

220 FAST FACT...

ALTHOUGH, movement is that of a very very slow snail's pace, Continental drift continues today, hence the Atlantic Ocean is slowly widening.

TECTONIC FORCE

THE COMPOSITION AND TEMPERATURE OF EARTH'S interior are the reasons nobody has ever gone there and probably never will. It's also what creates something called tectonic force.

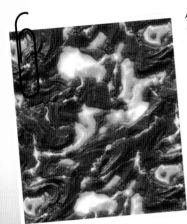

A closer look inside – although we stay nice and cool on the surface, beneath Earth's crust is the Mantle where temperatures reach nearly 3600°F causing the rock to have partly melted to form thick, molten rock called magma. Beneath the Mantle is Earth's core, which reaches temperatures in excess of 12600°F.

Altogether, this vast volume of incredibly hot stuff is a powerful source of pressure; tectonic force. But, what are the effects of tectonic force? Well, in short, it's mighty enough to create and rearrange continents, and in the process build mountains, cause earthquakes and volcanic eruptions to occur.

222 ## FAST FACT...

TECTONIC FORCE is the source of energy that drives plate tectonics.

223 MOVING MOUNTAINS

THE PLACE WHERE TWO PLATES MEET is called a plate margin and as plates move, stress and friction occurs along the plate margins. It's this stress and friction that is mostly responsible for the creation of fold mountains.

Along destructive plate boundaries, where plates move towards each other, the continental crust is squashed together and forced upwards. This process is called folding and it's this that creates fold mountains. This usually occurs where an oceanic plate is pushed underneath the continental plate but can also occur where two continental plates push towards each other. This is how mountain ranges such as the Himalayas and the Alps were formed.

Fold mountains are the most common type of mountain on Earth with others including volcanic mountains, erosional mountains (created as wind and water wear away soft portions of land and leave behind rocky hills) and fault-block mountains (created where parts of continental crust are displaced).

224 FAST FACT...

AT 29,035 FEET (8850 m), Mount Everest is the highest elevation on Earth.

225 FAST FACT...

THE HIMALAYAS which are known as fold mountains are growing by several centimeters each year due to the destructive margin where the Indian and Eurasian plates meet.

226 EARTHQUAKES

EARTHQUAKES ARE ALL TIED UP WITH PLATES that are slipping or scraping against each other and thus generating energy below the surface of the Earth. As this energy reaches the Earth's crust, the friction between the plates builds up and eventually is released by way of vibration of the Earth's surface. As the plates move, fractures in the Earth's crust, called fault lines, develop and earthquakes are often located on these.

227 FAST FACT...

THE RICHTER SCALE gauges how much energy was released at the earthquake's source. Each whole number on the Richter scale is a 10-fold increase over the previous number – Level 2 is barely detectable whilst Level 8 signifies total damage.

Most earthquakes are quite small and aren't readily felt. Larger and more violent ones, occurring when plates slide past or collide into one another can do tremendous damage though. An earthquake may last only a minute, but its affects can be devastating. Thousands of people died in 2010 for example, after a major earthquake struck Haiti.

228 FAST FACT...

THE SAN ANDREAS fault line between the Pacific and North American plates is a notorious Earthquake zone.

229 SEISMIC WAVES

🎓 **EARTHQUAKE ENERGY IS RELEASED IN SEISMIC WAVES.** These waves spread out from the focus, the point inside the Earth's crust where the pressure is released. The surface seismic waves that oscillate the Earth, either up and down or side to side are named, 'Love', after a famous mathematician and 'Rayleigh', after a famous physicist.

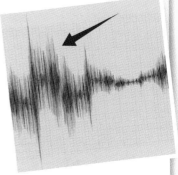

The point on the Earth's surface above the focus is called the epicenter, and that's where seismic waves are felt most strongly. The most severe damage will happen close to the epicenter, and as the waves become less strong as they travel further away, so the extent of any damage decreases.

When seismic waves reach the air, they take the form of sound waves, hence they travel at the speed of sound and then are often referred to as shock waves. Traveling through Earth though, seismic waves travel at different speeds and behave differently. Over the past several decades, listening devices have been placed within the Earth's crust that record and analyze seismic waves, providing along with other factors, information on what Earth may look like from the inside.

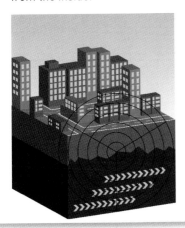

230 FAST FACT...

📖 **A SEISMOLOGIST** is someone who studies seismic waves and earthquakes.

231 FAST FACT...

📖 **SEISMIC SURVEYS,** which calculate the travel time of seismic waves, provided from three different locations in the World can be used to locate the epicenter of an earthquake.

VOLCANOES

🎓 **THEY MAY LOOK LIKE MOUNTAINS** but instead of being pushed up gradually, volcanoes form when part of Earth's inside, gets out. When mounting pressure comes to a peak, the interior literally erupts through the Earth's crust. In the most violent eruptions, huge black clouds of lava, ash, and dust spit out and molten lava races along the ground, burning everything in its path.

Volcanoes build up land surface, but eruptions can also be deadly. Mount Soufriere's eruptions during the 1990s for example, destroyed most homes on the Caribbean island of Montserrat. A 1985 volcanic eruption killed 23,000 people in Colombia and The Krakatoa eruption in Indonesia in 1883 killed over 36,000 people. The earthquake in AD 62 may have led to the eruption of Mount Vesuvius, one of the most famous disasters in history.

233 **FAST FACT...**

📖 **MOST OF THE WORLD'S VOLCANOES** are related to subduction (a scientific Latin word meaning 'carried under'), where an oceanic plate meets and slips under a continental plate.

Chains of underwater volcanoes have formed along the plate boundary, known as the Mid-Atlantic Ridge. One of these volcanoes may well become so large that it erupts out of the sea to form a volcanic island. Examples of other volcanic islands are Surtsey and the Westman Islands near Iceland.

234 **FAST FACT...**

📖 **VOLCANIC ERUPTIONS** can also trigger earthquakes and tsunamis.

235 MANAGING TECTONIC HAZARDS

🎓 **PREDICTION OF EARTHQUAKES AND VOLCANIC ERUPTIONS** can help minimize the damage caused. As a volcano becomes active for example, it gives off a number of warning signs, including hundreds of small earthquakes caused by magma rising up through cracks in the Earth's crust, a rise in temperature around the volcano and a release of gas containing sulphur, which signifies how imminent the eruption may be. These warning signs are picked up by volcanologists (experts who study volcanoes) who use monitoring techniques including thermal imaging and satellite cameras to detect heat around a volcano and chemical sensors are used to measure sulphur levels.

Although not as easy to predict as volcanic eruptions, laser beams can be used to detect plate movement, a seismometer can be used to pick up vibrations in the Earth's crust and levels of Radon Gas can be monitored to evaluate the chances of an earthquake.

236 FAST FACT...

📖 **VOLCANOES SUCH AS** Mount St Helens, USA and Mount Etna, Italy are closely monitored at all times because they've been active in recent years.

237 FAST FACT...

📖 **TRANSAMERICA PYRAMID** in San Francisco is a prime example of a building designed to absorb the energy of an earthquake and to withstand the movement of the Earth.

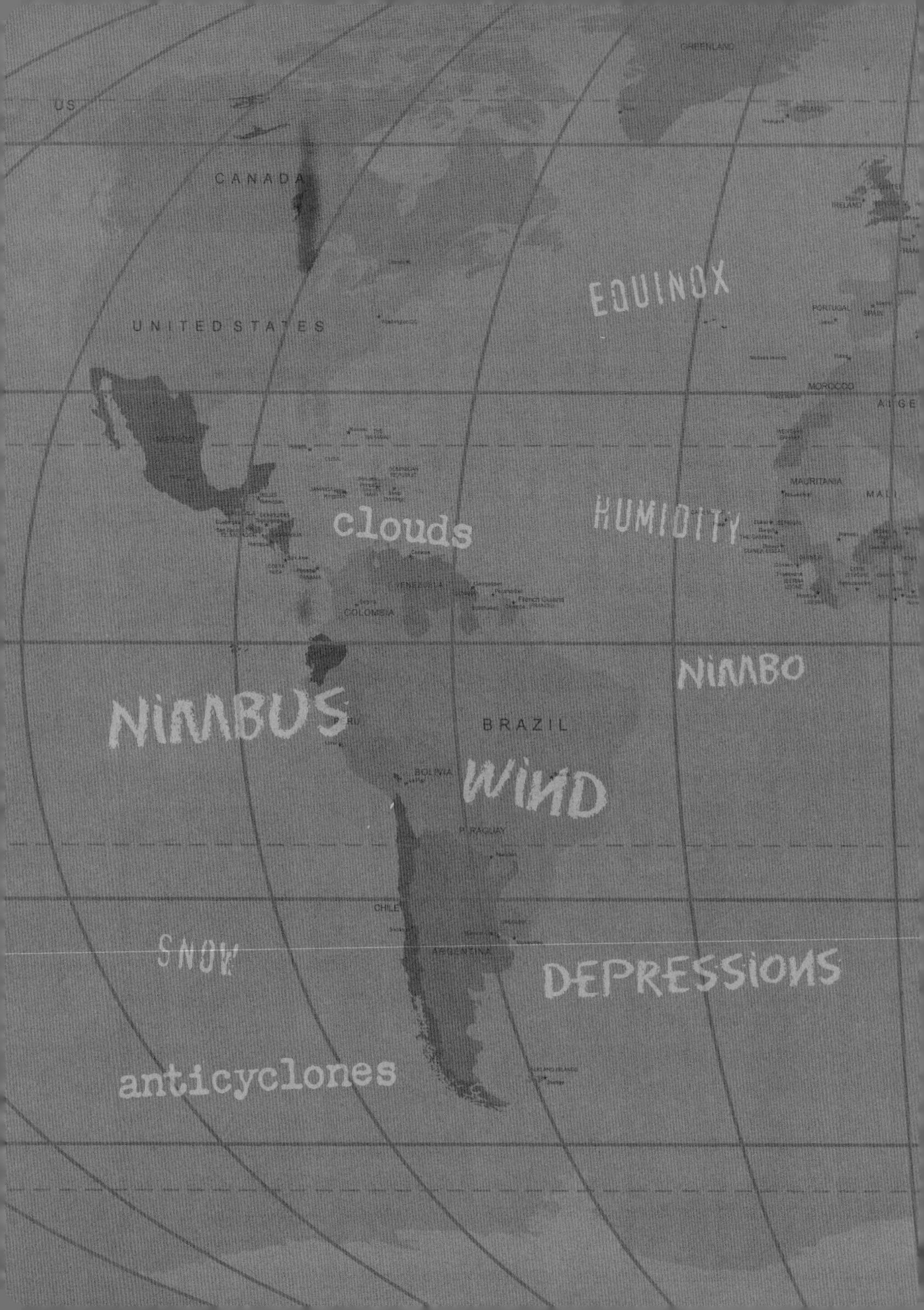
EQUINOX
clouds
HUMIDITY
NIMBO
NIMBUS
WIND
SNOW
DEPRESSIONS
anticyclones
US
CANADA
UNITED STATES
GREENLAND
IRELAND
PORTUGAL
SPAIN
MOROCCO
MAURITANIA
MALI
VENEZUELA
COLOMBIA
French Guiana
BRAZIL
BOLIVIA
PARAGUAY
CHILE
ARGENTINA

PRESSURE

JET STREAM

CORIOLIS

Weather & Climate

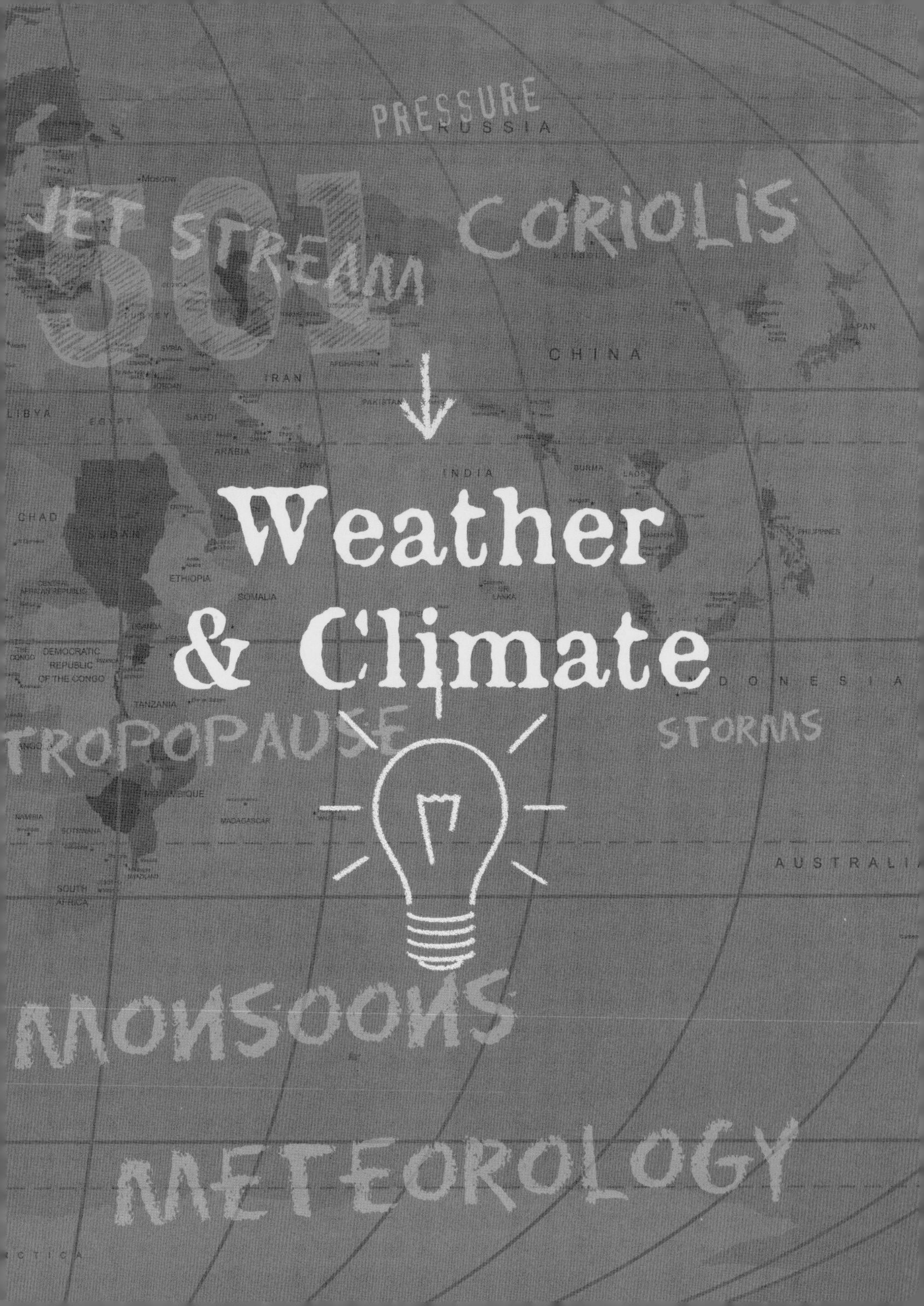

TROPOPAUSE

STORMS

MONSOONS

METEOROLOGY

WHAT CAUSES THE FOUR SEASONS?

🎓 **SINCE THE EARTH** is not only rotating around the Sun on a yearly cycle but is tilted on its axis at 23.5°, the Sun's rays strike the Earth at different points causing temperatures and daylight to change cyclically as well. We call these cyclical changes, spring, summer, autumn, and winter, the four seasons.

The cycles in the Earth's temperature and daylight, and therefore the seasons, are more distinct in certain parts of the Earth. The Tropic of Cancer, also referred to as the Northern tropic, is the circle of latitude on the Earth that marks the most northerly position at which the Sun may appear directly overhead and the Tropic of Capricorn, or Southern tropic, marks the most southerly latitude on the Earth at which the Sun can be directly overhead. This means that the changes in the seasons don't happen simultaneously, but rather occur at opposite times of the year, in the northern and southern hemispheres.

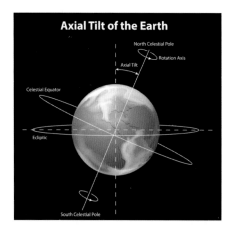

Axial Tilt of the Earth

North Celestial Pole
Rotation Axis
Axial Tilt
Celestial Equator
Ecliptic
South Celestial Pole

FAST FACT...

📖 **DURING AN EQUINOX,** occurring around March 21st and September 21st but not necessarily exactly on those dates, the rotation and tilt of the Earth causes the Sun's rays to hit the Equator directly.

240   HOW DO CLOUDS FORM?

🎓 **WE ALL KNOW** that clouds are needed before rain can fall. But how do clouds form in the first place? Well, the air is full of invisible water vapors, which condense and turn to liquid, or water droplets. These water droplets stick to salt and dust particles, known as aerosols. As the air rises, it cools and thus the ability to hold water vapor decreases and so condensation occurs. When the air is saturated and cannot hold any more water vapor, clouds form.

There are different types of clouds including, cirrus (very high wispy clouds), cumulo-nimbus (very deep storm clouds), cumulus (individual fluffy clouds), and stratus (flat gray layers of cloud, which can form at any height). Not all clouds means rain of course but the prefix or suffix, nimbus or nimbo signifies a rain-bearing cloud.

241   FAST FACT...

📖 **RELATIVE HUMIDITY** is the amount of water vapor in the air.

242   FAST FACT...

📖 **PRECIPITATION DOESN'T** just mean rain but also refers to hail, snow, fog, and dew. Hail and snow form when water droplets freeze, and fog and dew form when air is forced to cool close to the ground.

243 WIND — WHAT CAUSES IT?

WIND IS CAUSED BY DIFFERENCES in air pressure, and the greater the difference, the stronger the wind is. So, what causes air pressure to vary? Well temperature is the main factor. Warmer, lighter air is forced to rise creating areas of low pressure whereas heavier, colder air descends creating areas of high pressure. Height also plays a role because as the atmosphere becomes thinner, pressure naturally decreases.

Wind flows, as a horizontal movement of air across the Earth's surface, from high pressure (sinking) to low pressure (rising).

Sometimes seen as inconvenience perhaps, wind can have all kinds of effects on both human and physical geography. Considered one of four agents in erosion and deposition for example, wind even helps change the landscape of the Earth.

244 FAST FACT...

WIND DIRECTION is recorded using a wind vane and speed is recorded using an anemometer.

245 WEATHER FRONTS

WHERE TWO DIFFERENT air masses meet, a weather front occurs.

Warm fronts - formed when warm air rises over a mass of cold air. As the air lifts into regions of lower pressure, it expands, cools, and condenses the water vapor as wide, flat sheets of cloud.

Cold fronts - the transition zone where a cold air mass is replacing the warmer air mass. The cold air is following the warm air and gradually moves underneath the warmer air. When the warm air is pushed upwards it will rain heavily. Often more rain will fall in the few minutes the cold front passes than it will during the whole passage of a warm front. As the cold front passes, the clouds roll by and the air temperature is cooler.

Occluded fronts - occur at the point where a cold front takes over a warm front or the other way around. If a cold front undercuts a warm front it is known as a cold occlusion and if the cold front rises over the warm front it is called a warm occlusion.

246 FAST FACT...

OCCLUDED FRONTS bring changeable weather conditions.

DEPRESSIONS AND ANTICYCLONES

DEPRESSIONS ARE AREAS OF LOW atmospheric pressure formed when a warm air mass meets a cold air mass. Anticyclones, areas of high atmospheric pressure where the air is sinking, are the opposite of depressions. The creation of weather fronts (occurring where two different air masses meet), means that depressions produce cloudy, rainy, and windy weather whereas anticyclones are associated with settled, dry, and bright conditions.

248 FAST FACT...

AIR PRESSURE is the weight of the atmosphere at the Earth's surface. On maps, the differences in pressure are shown by isobars, which are lines joining points of equal pressure.

These conditions are due to the warming of the air as it sinks and the subsequent lack of clouds or rain forming. In summer therefore, an anticyclone can mean dry hot weather but in winter, the associated clear skies can mean cold, frosty nights. In cold conditions, because the cold forces moisture in the air to condense at low altitudes, anticyclones may also bring fog and mist.

250 FAST FACT...

THE INWARD, anticlockwise blowing winds of a depression are known as the Coriolis Force.

249 FAST FACT...

DEPRESSIONS often begin in the North Atlantic where warm, moist tropical air moving north, meets cold polar air moving south.

251 JET STREAMS

A JET STREAM is a fast-flowing narrow air current found near boundaries of adjacent atmospheric air masses with significant differences in temperature. The main jet streams are located in the tropopause (the boundary in the atmosphere between the troposphere and stratosphere).

Famously used by jet planes, the jet stream blows in a wave shape (known as the Rossby Wave), accelerating, and sucking air up from the troposphere and producing a low pressure cell as it moves from north to south and vice versa, decelerating, creating high pressure, as it turns. Jet streams may start, stop, split into two or more parts, combine into one stream, or flow in various directions including the opposite direction to most of the current.

But, what are they caused by? Jet streams are caused by a combination of a planet's rotation on its axis and atmospheric heating (solar radiation).

252 FAST FACT...

A JET STREAM is one of very few examples of when we have to look higher, beyond the troposphere, to get an explanation of a weather condition.

253 FAST FACT...

JET STREAMS blow incredibly quickly, up to nearly 25 miles an hour. Polar jets are the strongest, blowing at around 4-8 miles (23,000–39,000 ft) above sea level, and the higher and somewhat weaker subtropical jets at around 6–10 miles (33,000–52,000 ft).

254   TROPICAL STORMS

A TROPICAL STORM is a large low pressure system that develops over oceans in the Tropics during the summer. They move north and south away from the Equator along similar tracks and are known in different parts of the world as cyclones, hurricanes, and typhoons.

A tropical storm is hazardous, particularly if it reaches land as not only can the associated high winds (often exceeding 125 mph) destroy whole communities, buildings and communication networks but they can generate abnormally high waves and tidal surges. In fact, sometimes the most destructive elements of a tropical storm are the subsequent high seas and flooding making low-lying coastal areas especially vulnerable. The high levels of rainfall can also cause flooding and landslides of course.

255   FAST FACT...

TROPICAL STORMS can last as long as a month and although they travel very slowly - usually at about 15 mph (24 km/h) - wind speeds can reach over 75 mph (120 km/h).

256 HOW ARE TROPICAL STORMS FORMED?

WHEN WARM WET AIR, typically found over tropical oceans that have a water temperature of at least 80°F, rises, it condenses to form towering clouds and heavy rainfall. The rising warm air also causes pressure to decrease at higher altitudes. Warm air is under a higher pressure than cold air though, so moves towards the space occupied by the colder, lower pressure air, and subsequently causes the low pressure to suck in air from the warm surroundings, which then also rises.   A continuous upflow of warm and wet air continues to create clouds and rain.

The warm wet air also creates a low pressure zone near the surface of the water and the air around it begins to flow in a spiral at very high wind speeds. This causes the cycle to continue further with air being ejected at the top of the storm (can be up to 9 miles high), and falling away from the eye of the storm. The mass of air over the eye of the storm is reduced causing the wind speed to increase further. The faster the winds blow, the lower the air pressure in the center, and so the cycle continues causing the storm to grow stronger and stronger.

257 FAST FACT...

📖 **THE EYE OF A STORM** forms because this is the only part where cold air is descending.

258 FAST FACT...

📖 **AS TROPICAL STORMS** move inshore, their power gradually reduces because their energy comes from sucking up warm, moist sea air.

259 MONSOONS

EVERY SUMMER, SOUTHERN ASIA AND especially India, is drenched by rain that comes from moist air masses that move in from the Indian Ocean to the south. These rains, and the air masses that bring them, are known as monsoons.

However, the term monsoon refers not only to the summer rains but to the entire cycle that consists of both summer moist onshore winds and rain from the south as well as the offshore dry winter winds that blow from the continent to the Indian Ocean.

But, what causes a Monsoon? The precise cause is not fully understood, but we do know that air pressure is one of the primary factors. In the summer, a high pressure area lies over the Indian Ocean while a low exists over the Asian continent. The air masses move from the high pressure over the ocean to the low over the continent, bringing moisture-laden air to south Asia. During winter, the process is reversed and a low sits over the Indian Ocean while a high lies over the Tibetan plateau so air flows down the Himalayas and south to the ocean.

Smaller monsoons take place in equatorial Africa, northern Australia, and, to a lesser extent, in the southwestern United States.

260 FAST FACT...

THE ARABIC WORD for season, mausim, is the origin of the word monsoon due to their annual occurrences.

261 FAST FACT...

ALMOST HALF OF THE WORLD'S population lives in areas affected by the monsoons of Asia and most of these people are subsistence farmers, so too much or two little rain can mean disaster in the form of famine or flood.

262 FAST FACT...

MONSOONS ARE ESPECIALLY important to India, Bangladesh, and Myanmar (Burma) as they're responsible for almost 90 % of India's water supply.

TORNADOES

A TORNADO IS A VIOLENTLY rotating column of air that is in contact with both the surface of the Earth and a cumulonimbus cloud or, in rare cases, the base of a cumulus cloud. They are often referred to as twisters or cyclones, although the word cyclone is used in meteorology, in a wider sense, to name any closed low pressure circulation.

FAST FACT...

TORNADOES CAN BE detected before or as they occur through the use of Pulse-Doppler radar.

FAST FACT...

THERE ARE SEVERAL SCALES, including The Fujita scale and the TORRO scale, for rating the strength of tornadoes.

How can we spot a tornado? Tornadoes come in many shapes and sizes, but are typically in the form of a visible condensation funnel, whose narrow end touches the Earth and is often encircled by a cloud of debris and dust. Even if you didn't spot a tornado, you'd feel one. Most tornadoes have wind speeds less than 110 miles per hour (177 kms/h) but the most extreme ones can attain wind speeds of more than 300 miles per hour (483 kms/h).

Tornadoes have been observed on every continent except Antarctica. However, the vast majority of tornadoes in the world occur in the Tornado Alley region, i.e., the area between the Rocky Mountains and Appalachian Mountains in the United States.

266 METEOROLOGY

METEOROLOGY IS THE SCIENCE of monitoring and studying the atmosphere and predicting its weather and climate. Weather and climate are two different things of course. Weather describes the atmospheric conditions at a particular place and time together with the changes taking place over the short term (hour-by-hour or day-by-day). Climate on the other hand refers to what is expected to happen in the atmosphere rather than the actual conditions. It's possible for the weather to be different from that suggested by the climate.

Recording of data is key to meteorology and at a weather station all the instruments have precise locations so that the data gathered is reliable and comparable. Not just at weather stations but automatic instruments also exist in remote locations (e.g., ocean buoys), ships and airplanes report weather conditions as do satellites and radar images. Readings are taken at least once every 24 hours and the information is plotted on maps or synoptic charts and then analyzed to obtain a weather forecast.

267 FAST FACT...

METEOROLOGISTS are the people responsible for recording and analyzing weather.

WHAT EFFECTS CLIMATE?

CLIMATE IS AFFECTED by many factors including latitude, altitude, distance from the sea, prevailing winds, and ocean currents.

Latitude – the further away from the Equator, the lower the temperature becomes. This is because solar energy (insolation) is less concentrated towards the Poles.

Altitude – locations at a higher altitude (height above sea level) have colder temperatures. On average, temperature decreases by 34°F for every 100 meters in altitude.

269 FAST FACT...

INFLUENCING vegetation and wildlife as well as human activity, climate is fundamental to life on Earth.

Distance from the sea – oceans heat up and cool down much more slowly than land. This means that coastal locations tend to be cooler in summer and warmer in winter than places inland at the same latitude and altitude.

Prevailing wind – this is the most frequent wind direction a location experiences.

Ocean currents – warm and cold currents move large distances across oceans. For example, a warm ocean current called the North Atlantic Drift keeps Britain warmer and wetter than places in continental Europe.

270 FAST FACT...

GLOBAL CLIMATE zones with similar flora, fauna, and climate are called biomes.

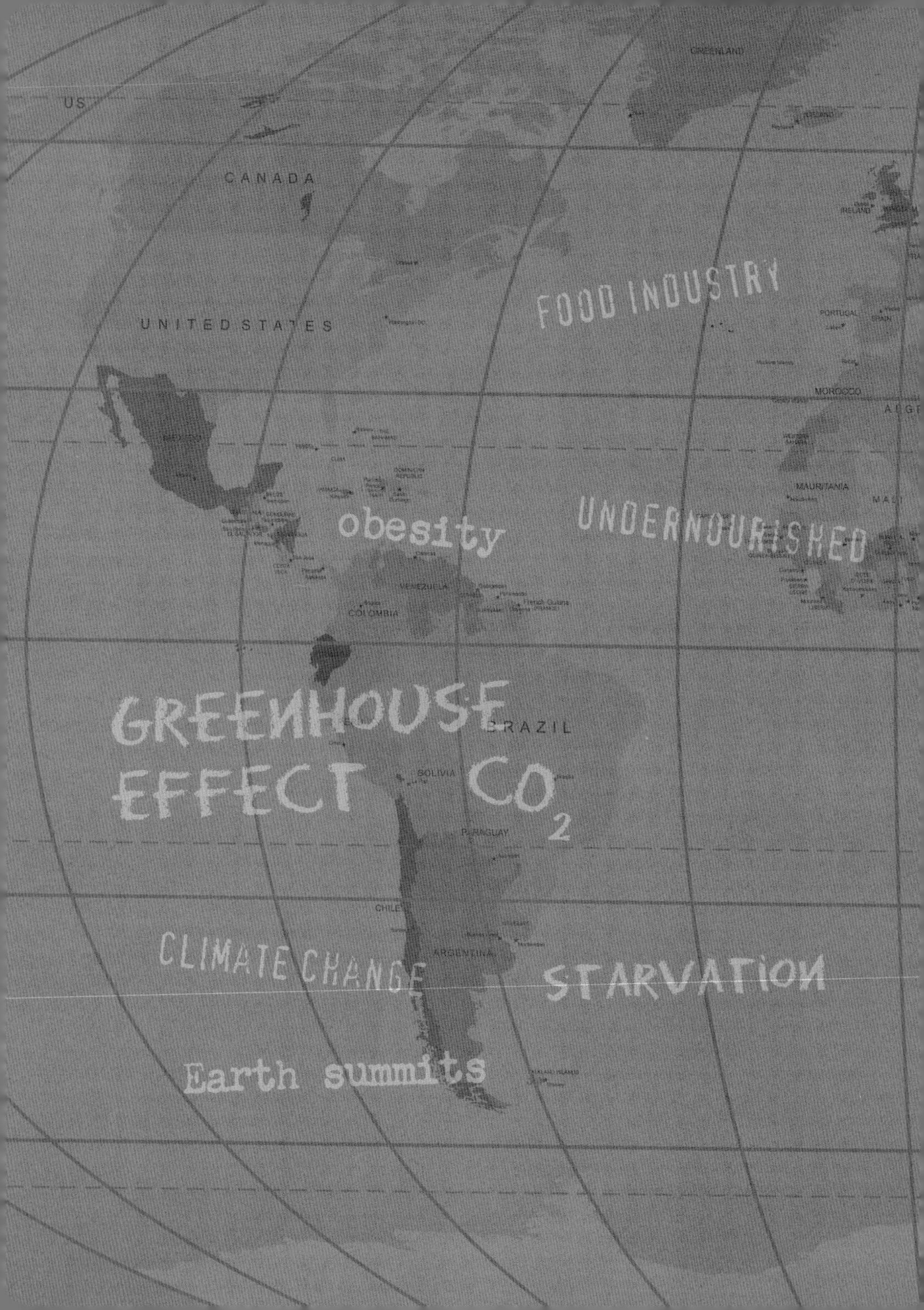

GREENLAND
US
CANADA
UNITED STATES
FOOD INDUSTRY
PORTUGAL
MOROCCO
MAURITANIA
obesity
UNDERNOURISHED
VENEZUELA
COLOMBIA
GREENHOUSE EFFECT CO_2
BRAZIL
BOLIVIA
PARAGUAY
CHILE
ARGENTINA
CLIMATE CHANGE
STARVATION
Earth summits

POLLUTION

RUSSIA

501

ALTERNATIVE

ENERGY

CHINA

IRAN

PAKISTAN

LIBYA

EGYPT

SAUDI
ARABIA

CHAD

SUDAN

ETHIOPIA

CENTRAL
AFRICAN REPUBLIC

GLOBAL
ISSUES

PHILIPPINES

THE
CONGO

DEMOCRATIC
REPUBLIC
OF THE CONGO

TANZANIA

INDONESIA

DESALINIZATION

NAMIBIA

BOTSWANA

MADAGASCAR

MOZAMBIQUE

SOUTH
AFRICA

AUSTRALIA

Methane

WATER SUPPLY

AGRICULTURE

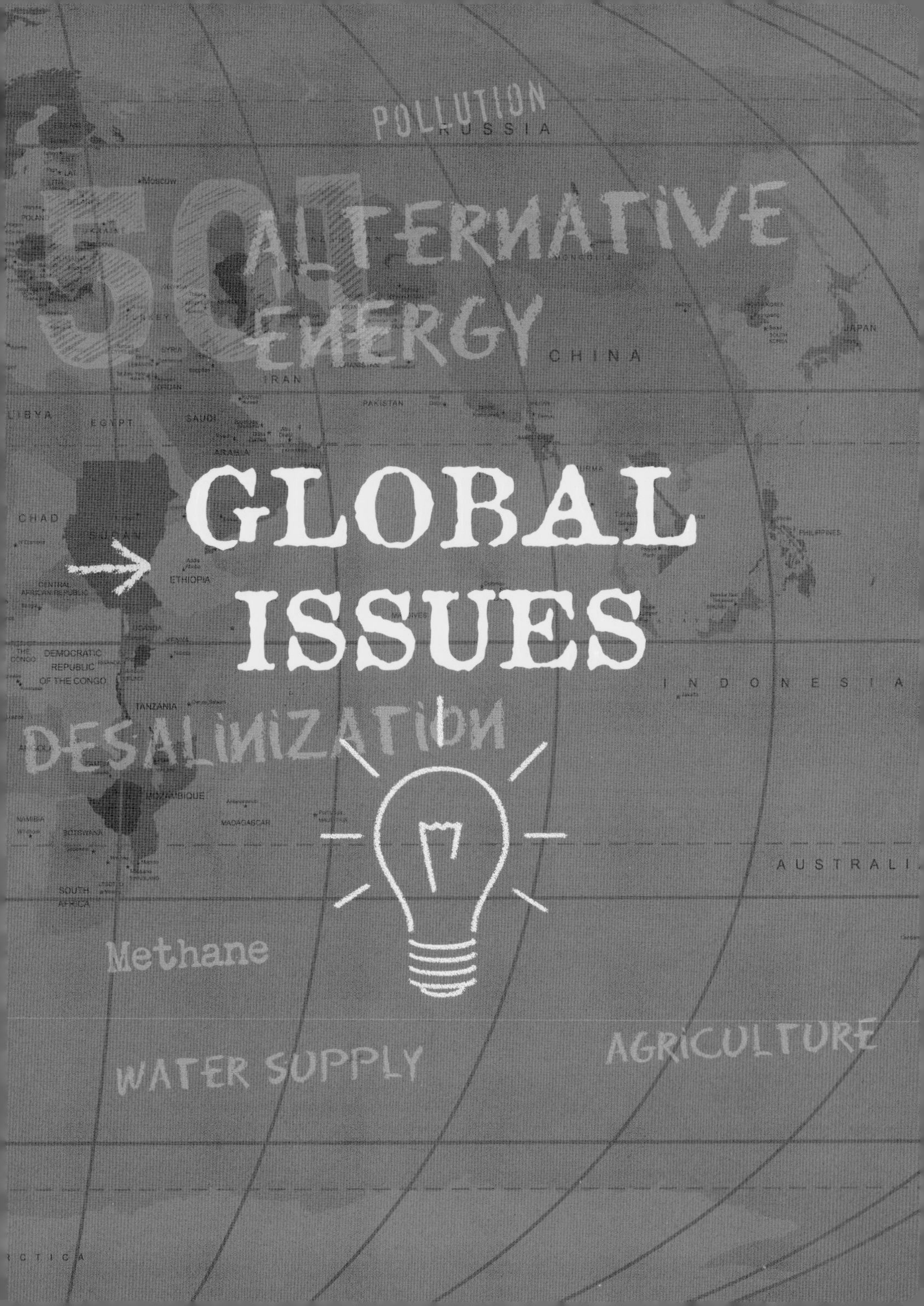

271 GLOBAL FOOD INDUSTRY

🎓 **GLOBALIZATION HAS MADE THE FOOD-SUPPLY** chain a complex one, involving more countries than ever before. For example, rice grown in Thailand may be packaged in India and sold in a supermarket in the UK.

There is also a growing issue of food shortages in some countries, and as demand increases whole nations are having to become more adept at managing food resources.

Globally, food is in more demand for many reasons including a growing population (the global population increases by 75 million each year), increased income (therefore people can afford foods like meat and dairy) and more extreme weather conditions (droughts and extreme rainfall ruin crops so less can be sold on the global food market).

At a global scale, food consumption is shockingly uneven. Some more economically developed countries have problems relating to the over consumption of food and suffer instead from obesity and heart disease. Meanwhile, in less economically developed countries, under nutrition is a very real problem.

272 FAST FACT...

📖 **76%** of the population in The Democratic Republic of Congo are undernourished due to lack of food supply and growing demand.

273 WHAT EFFECTS CLIMATE?

🎓 **WITH 70% OF EARTH'S** surface being covered by water, you'd probably expect that there's more than enough supply. The vast majority of Earth's water is unavailable for human use though.

In some countries there are desalinization plants that, as the name implies, remove salt from seawater and produce fresh water. The process is expensive though and helps satisfy the needs of only a few localities, mainly coastal cities.

Not only is the desalinization process limiting when it comes to global water supply but also the geography of supply is very uneven according to the geography of need, i.e, fresh water is often abundant in areas where human need for it is scarce, and conversely, fresh water is often scarce where human need is abundant.

Although the total fresh water supply is not used up, much has become polluted, salted, unsuitable or otherwise unavailable for drinking, industry, and agriculture.

274 FAST FACT...

📖 **¾ OF THE FRESH WATER** on Earth is solid, existing in ice sheets.

275 FAST FACT...

📖 **A WATER CRISIS** is a situation where the available potable (safe enough for human consumption), unpolluted water within a region is less than that region's demand. According to the United Nations, by 2025, two-thirds of the world population could be in water crisis.

276 GREENHOUSE EFFECT

GREENHOUSE GASES, atmospheric gases such as carbon dioxide (CO_2) and methane that exist to keep our planet warm are enhanced by industrial processes and agriculture, and are having a global impact on temperature and weather systems. Scientists believe that the build-up of so-called greenhouse gases in the atmosphere acts like a blanket or greenhouse around the planet; heat is trapped inside the Earth's atmosphere. This is the greenhouse effect, and the resulting increase in global temperatures is called global warming.

How the greenhouse effect works - As human activity such as the burning of fossil fuels and deforestation increases, greenhouse gases are released. Normally, when heat enters the atmosphere, it is through short-wave radiation; a type of radiation that passes smoothly through our atmosphere. As this radiation heats the earth's surface, it escapes the earth in the form of long-wave radiation; a type of radiation that is much more difficult to pass through the atmosphere. Greenhouse gases released into the atmosphere cause this long-wave radiation to increase. Thus, heat is trapped and creates a warming effect.

277 FAST FACT...

THE BIGGEST PRODUCERS of CO_2 in the world are the United States (USA), China, Russia, Japan, India, Germany, United Kingdom, Canada, Italy, and Mexico.

278 FAST FACT...

A GROUP OF GREENHOUSE gases called chlorofluorocarbons (CFCs) have been responsible for depleting the ozone layer as they attack and destroy ozone molecules.

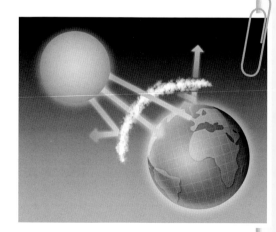

279 CLIMATE CHANGE — DIRECT EFFECTS

🎓 **OUR CLIMATE IS CONTINUALLY** changing but since about 1950, records show that there's been a steep climb in global temperature compared to the past. As Earth's climate changes, the polar regions are warming and much of the icepack covering the Arctic Ocean melts during the summer months, for example.

Put simply, global warming is bad news for our planet. Here are some of the direct effects:

Agriculture –

- Agricultural land on the edge of deserts becomes unusable, through the process of desertification.
- Crop yields are expected to decrease, and prices to rise for all major world crops.
- Crops could be wiped out in low-lying areas that suffer from flooding.

Change in Sea Levels –

- Coastal land is at risk, especially land on deltas.
- Sea defences are under more stress
- Low-lying land is threatened.

Water and Ice –

- More mass movement may occur as glaciers melt.
- Less fresh water will be available in coastal areas as it will mix with seawater.

280 FAST FACT...

📖 **SOME SCIENTISTS ESTIMATE** that atmospheric temperatures could rise by 34.5°F – 42.4°F in the next 100 years.

281 FAST FACT...

📖 **GREATER AMOUNTS OF ENERGY** in the atmosphere will increase the likelihood of extreme events such as tropical storms.

282 CLIMATE CHANGE — KNOCK-ON EFFECTS

CLIMATE CHANGE also has many projected knock-on effects for us, the people of this planet.

- As the world population increases, more people will be living in cities located on the coast or low-lying land, and as a result, will be threatened by flooding.
- People will migrate from areas suffering drought. Any that remain will be in danger of dying from starvation and lack of water.
- Wide-spread crop failures will result in malnutrition or starvation in central Africa.
- Communities, especially in Asia, that use the melt water from glaciers may see this supply decrease.
- Economies that rely on skiing as a form of income may suffer as the skiing season is reduced or disappears through lack of snow.

283 FAST FACT...

17 MILLION PEOPLE in Bangladesh alone will be threatened by flooding due to climate change.

284 FAST FACT...

THE SUBMERSION OF ISLANDS in the Caribbean and Indian Ocean will produce many millions of refugees.

285 STRATEGIES TO REDUCE GLOBAL WARMING

HUMAN ACTIVITIES have an impact on global warming. Depending on where an individual lives average emissions of carbon dioxide will vary with those living in an economically developed country having higher emissions, known as carbon footprint, than those in less developed countries. The world average is about 4 tons of carbon dioxide per person per year. Whilst individuals can reduce carbon footprint by doing simple things like taking the bus instead of the car or switching lights off when they aren't needed, global strategies are in place to try and reduce global warming.

For example, alternative forms of energy such as solar, tidal, wind, and hydroelectricity are being developed. A switch to nuclear power is being considered, although views on the future of this are divided with those against it pointing to the accidents of the past and the problems of nuclear waste disposal. Alternative energy sources for road transport (i.e, electric cars) are also being developed.

286 FAST FACT...

EARTH SUMMITS exist to try and reduce emissions and limit pollution on a global scale.

ALTERNATIVE ENERGY — OCEAN CURRENTS

TODAY, OCEAN CURRENTS are gaining significance as sources of alternative energy. Because water is dense, it carries an enormous amount of energy that could possibly be captured and converted into a usable form through the use of water turbines. Currently this is an experimental technology being tested by the United States, Japan, China, and some European Union countries.

288 FAST FACT...

📖 EXISTING TECHNOLOGIES

require an average current of five or six knots to operate efficiently, while most of the Earth's currents are slower than three knots.

Whether ocean currents are used as alternative energy, to reduce shipping costs, or in their natural state to move species and weather worldwide, they are significant to geographers, meteorologists, and other scientists because they have a tremendous impact on the globe and earth-atmosphere relations.

But how are ocean currents utilized for energy? Well, there are currently various types of turbines and wave generators but a new device, inspired by the way fish swim, consists of a system of cylinders positioned horizontal to the water flow and attached to springs. As water flows past, the cylinder creates vortices, which push and pull the cylinder up and down. The mechanical energy in the vibrations is then converted into electricity. Cylinders arranged over a cubic meter of the sea or river bed in a flow of three knots can produce 51 watts.

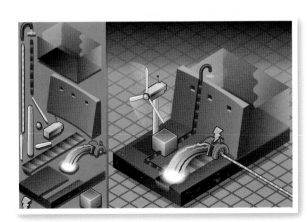

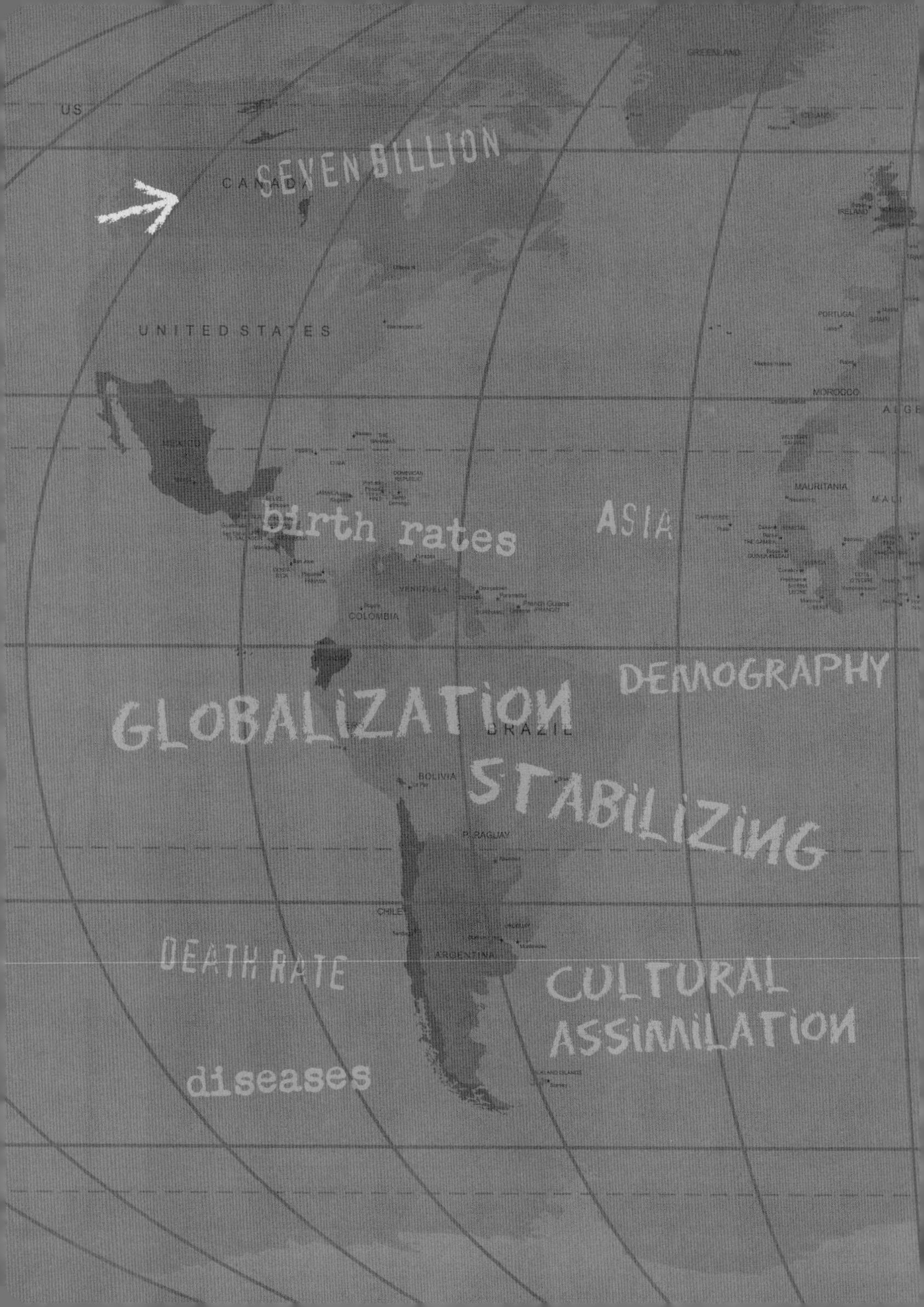
SEVEN BILLION
birth rates
ASIA
DEMOGRAPHY
GLOBALIZATION
STABILIZING
DEATH RATE
CULTURAL ASSIMILATION
diseases
US
CANADA
GREENLAND
IRELAND
PORTUGAL
SPAIN
MOROCCO
UNITED STATES
MEXICO
MAURITANIA
MALI
VENEZUELA
COLOMBIA
BRAZIL
BOLIVIA
PARAGUAY
CHILE
ARGENTINA

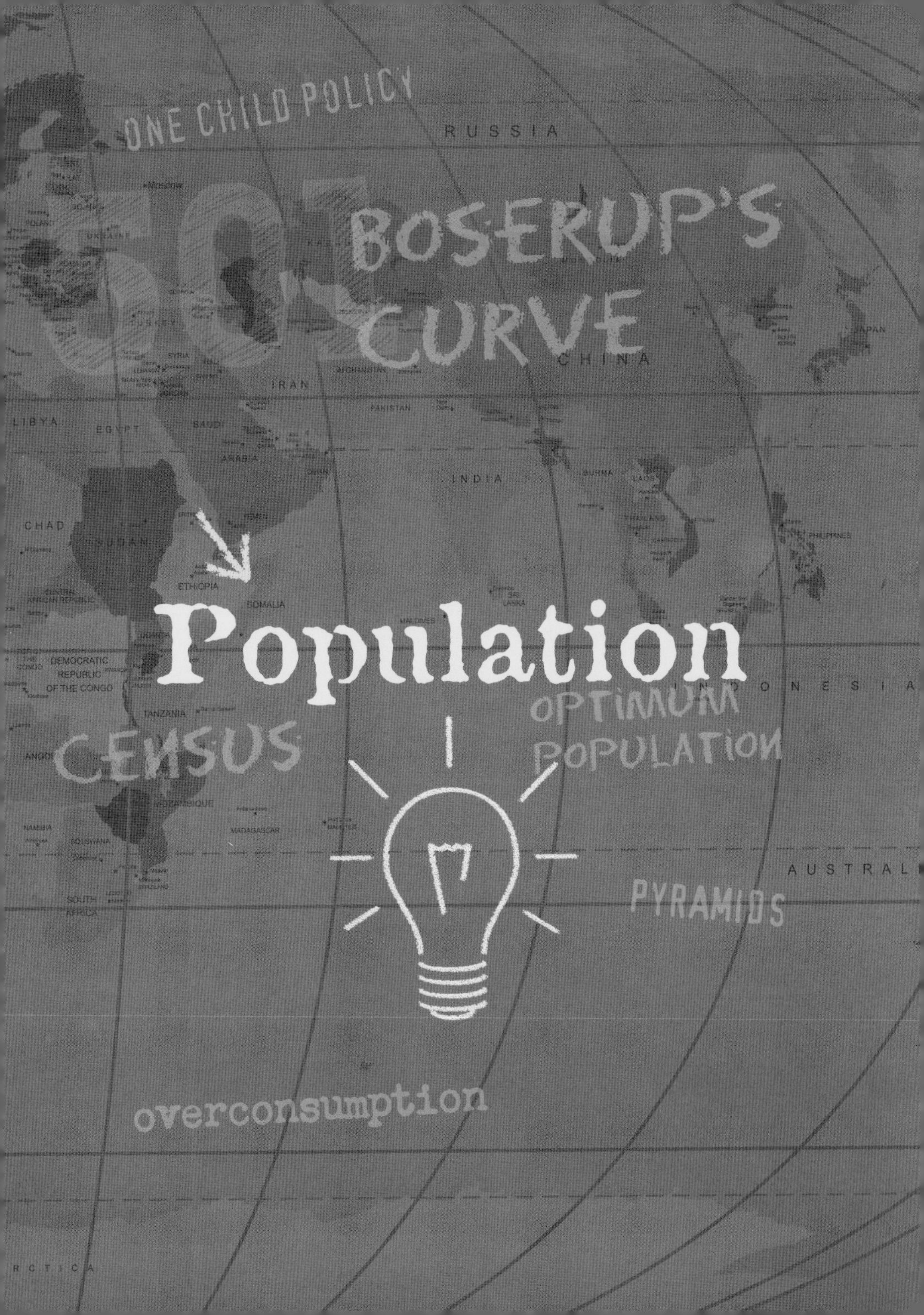

ONE CHILD POLICY

RUSSIA

BOSERUP'S
CURVE

CHINA

Population

OPTIMUM
POPULATION

CENSUS

PYRAMIDS

AUSTRAL

overconsumption

WORLD POPULATION

WORLD POPULATION is defined as the number of human beings currently living on Earth. In March 2012, it was estimated to be over seven billion.

The population has been increasing since the end of the bubonic plague the Black Death in 1350, when the total number of people was around 370 million. Since then it reached one billion in 1804, two billion in 1927 and then accelerating rapidly to three billion in 1960, four billion in 1974, five billion in 1987 and six billion in 1999.

Over half the world's population, 60% live in Asia, which is home to the most populous country, China, and city, Tokyo. The most densely populated countries are in Asia as well, with Bangladesh home to more people than Russia – 160 million compared to 142 million – even though Bangladesh is around 56,977 sq mi and Russia is 6,592,800 sq mi.

Predictions vary for the future of the global population, with some demographers proposing the population will only increase to 7.5 billion by 2050 and others predicting an increase to 10.5 billion.

290 FAST FACT...

THE CONTINENT with the smallest population is Antarctica, which varies from around 4,500 in summer to 1,100 in winter at its research stations.

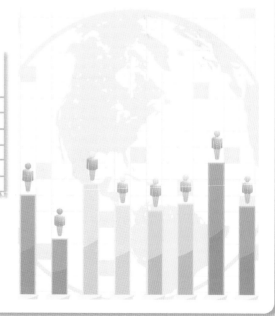

291 POPULATION DISTRIBUTION

DESPITE OVER SEVEN BILLION PEOPLE living on just less than 57.3 million square miles (149 million square kilometers) of land worldwide, there are tremendous differences in density, due to Population Distribution.

On a global level there are many reasons for this, including land relief – flat, low-lying areas such as river deltas are likely to have better soil than high-altitude mountainous ones – fresh water supplies, climate, vegetation, soils, disease and pests, political factors such as investment, communications passages such as rivers that allow and natural resources.

292 FAST FACT...

ANTARCTICA has the second largest land area after Russia with 6,592,800 sq mi, but only 1-2,000 people live there.

For national population distribution, there will be a combination of physical and human factors again, although there may be more anomalies as expected with a smaller sample. For example, the population of Brazil is mostly distributed along the coast, however large settlements like Manaus exist in the heart of the Amazonian jungle as it built up around the rubber industry located there.

293 FAST FACT...

PERTH, AUSTRALIA, is surprisingly remote: the nearest city with a population of 10,000 or more is Adelaide 1,300 miles (2,100 kms) away.

BIRTH RATES are a way of representing the increase in the human population for a particular country or area, usually presented as a ratio of births per thousand members of the population per year.

Many factors can affect birth rates, including government population policies, family planning services and contraception, mortality rates and social, and religious beliefs. In some countries, birth rate can become a national issue as with China's one child policy, which has been controversial with its implication in forced abortions, female infanticide and underreporting of female births.

295   FAST FACT...

THE COUNTRY with the highest birth rate is Niger in Africa, with 51.26 births per thousand people and the country with the lowest birth rate in Japan with 7.64 births per thousand.

Since the 1950s, the world crude birth rate is dropped from 37.2 per thousand to around 20 per thousand, creating the problem of an aging population. In Japan in 1950 there were 9.3 people under 20 for every person over 65, by 2025 this is forecast to be 0.59 people under 20 for every person over 65.

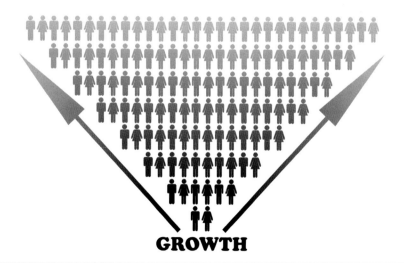

GROWTH

DEMOGRAPHY

🎓 **THE STATISTICAL STUDY OF A POPULATION** is called demography and helps to provide an idea of size, structure, and distribution of a group of people.

There are a wide variety of changes in a population that can be studied, including birth, migration, aging and death, and the study of demographics dates back to ancient societies in Greece, Rome, India, and China.

Demography can help to provide governments, companies, and individuals with an idea of general fertility rates as well as infant mortality and the average age of a population that enables forward planning and effective population management techniques.

Demographic data collection can be carried out using direct methods, where registry statistics and censuses are used to accurately gather complete data, and indirect methods when full data is not available. These often use information about brothers and sisters to extrapolate data and provide estimates.

297 FAST FACT...

📖 **DEMOGRAPHY COMES** from the ancient Greek 'demo' meaning 'the people' and 'graphy' meaning 'measurement.'

298 DEMOGRAPHIC TRANSITION MODEL

THE DEMOGRAPHIC TRANSITION MODEL describes the change in a country's birth and death rates as it develops from a pre-industrial to industrialized economic system.

The theory was first introduced as a model by the American demographer Warren Thompson in 1919, who had observed changes in the more developed countries over the previous 200 years. The model itself involves four stages, with additional stages proposed.

Stage one occurs when birth and death rates are both relatively high and the population stay constant with a fairly young demographic.

Stage two is a fall in death rates and increasing population, brought on by increasing food supplies and public health improvements through sanitation and personal hygiene.

Stage Three is a stabilizing of the population as birth rates fall. This can be because parents realize their children have a better life expectancy, because of the cost of educating children, increasing female employment, and improved access to birth control.

Stage four is low birth and death rates with a high stable population.

Stage five has been proposed as the decline of an aging population as some European and East Asian countries have sub replacement fertility with higher death rates than birth rates.

299 FAST FACT...

SOME COUNTRIES such as Brazil and China have passed through the model rapidly due to fast economic and social change while some countries in Africa have stalled at stage two.

300 PESTS AND DISEASES

🎓 **THROUGHOUT THE HISTORY OF HUMANS**, one of the biggest factors affecting population size has been Pests and Diseases.

The world population has been increasing since the 14th century, but only after the Black Death swept through Asia and Europe, killing an estimated 75-200 million people at a time when there were only 450 million on the planet.

As well as reoccurrences of the bubonic plague throughout the subsequent centuries, there have been many other instances of disease affecting population, as well as populations affecting disease. In Southern Italy there was a long standing history of malaria because of the numerous swamps providing an excellent environment for mosquitoes carrying the disease. Land reclamation leads to the swamps being drained and a fall in the number of malaria cases for the region in the early 20th century.

301 FAST FACT...

📖 **IN EGYPT, BILHARZIA** snails are a major risk to human health. The freshwater snails carry schistosomiasis and cause an average nine deaths everyday.

302 MANAGING POPULATION CHANGE

AS THE POPULATION DEMOGRAPHIC changes worldwide, individual countries have developed strategies to cope with foreseeable problems. This is called Managing Population Change.

For LEDCs (Least Economically Developed Countries), the key areas of concern are rapid population growth due to a high birth rate. For MEDCs (More Economically Developed Countries), the major issue is an aging population as they go through the latter stages of the Demographic Transition Model where the birth rate and death rate has declined. This creates a society with fewer people of working age supporting an elderly population.

Governments around the world have tried to counter these problems with initiatives such as China's One Child Policy. Begun in the 1970s, this policy imposes fines on any couple that have more than one child. In other countries such as Italy, which had the fastest aging population in Europe and the lowest birth rate, cash incentives were introduced in the 2000s for couples having a second child.

303 FAST FACT...

SINCE IT WAS INTRODUCED in the 1970s, Chinese authorities claim the One Child Policy has prevented more than 400 million births.

304 BOSERUP'S CURVE

THE WORLD POPULATION has been increasing since the 14th century as industrialization has lead to a fall in the crude death rate. At the end of the 18th century, British scholar Malthus plotted graphs to show his theory of population growth.

According to Malthus, population grows faster than food supply and therefore leads to two possible outcomes, either the population is reduced by famine, war and disease or the population could slow its growth to stay within the boundaries of food production.

However, Danish economist Ester Boserup argued that because necessity was the mother of invention, food supply would always increase to meet demand as new technology developed to meet requirements. This can be represented on a graph plotting quantity against time, with the food supply line plotted above the population line, both curving upwards but never intersecting.

There are compelling arguments for both sides; there is enough food to feed the world, assisted by the Green Revolution producing seeds that increased yields eight-fold, but there may be future crises such as global water shortages that may preclude this.

305 FAST FACT...

BOSERUP WAS A believer in humanity and said: "The power of ingenuity would always outmatch that of demand".

306 CENSUS

AS FAR BACK AS THERE HAS BEEN ORGANIZED society, leaders have sought to record the number of people in their community, as well as some general details about their age, gender, and activity. The word census comes from the Latin 'censere' meaning 'to estimate' and was first used by the Romans as they kept track of men fit for military service.

There are records of censuses being conducted by the Babylonians almost 6,000 years ago and it appears that every six or seven years they would record the number of people and livestock plus the reserves of honey, butter, vegetables, milk, and wool.

In medieval Europe, William the Conqueror carried out a census known as the Domesday Book in 1086 to establish taxes for the land he had taken control of.

A modern census is carried out in most countries to gather data for research, marketing, planning, and to assist political representation.

307 FAST FACT...

THE OLDEST EXISTING census was taken in China in the year 2 AD. It recorded the population at 59.6 million and is considered quite accurate.

308 FAST FACT...

THE INCA EMPIRE did not have a written language so instead used knots on cords made of alpaca or llama hair to record values based on a positional system.

309 OPTIMUM POPULATION

🎓 **OPTIMUM POPULATION** is a term used to describe an ideal size of a community for a specific area. It can be used for a small settlement like a town or village through to a country or even the entire planet.

There are differing opinions on what 'optimal' means, but it usually takes into account several factors such as ecological sustainability, basic human rights, a certain degree of resources and wealth, and the preservation of culture.

In order to reach an optimum population, there must already be the apparatus in place to measure human welfare and environmental impact, as well as policies to control birth rate and migration.

Critics of the optimum population concept argue that the human population will ultimately be able to adapt to any increase in numbers, and that the necessary assumptions make the concept speculative.

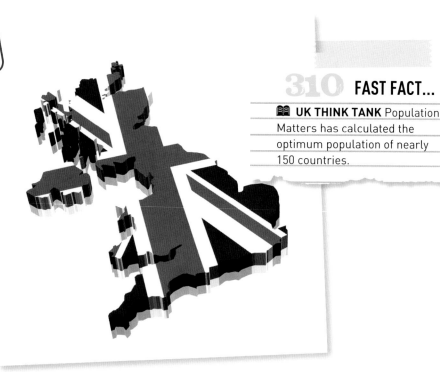

310 FAST FACT...

📖 **UK THINK TANK** Population Matters has calculated the optimum population of nearly 150 countries.

OVERCONSUMPTION – OF THE POPULATION

🎓 **OVERCONSUMPTION** occurs when the demand for natural resources has grown to exceed their availability from the ecosystem. The concept of overconsumption is closely tied to overpopulation, as the more people there are the greater their consumption of raw materials.

The main cause of overconsumption is consumerism and the idea of always needing another product or item, such as children's toys. Once the child grows out of the toy or breaks it, it is discarded and frequently ends up in landfill sites. Manufacturers also build in planned obsolescence – designing products that wear out or become outmoded after limited use.

As well as straining resources, overconsumption produces huge amounts of waste, leading to phenomena like the North Pacific Garbage Patch – and area of waste plastic and chemicals with size of Texas floating in the Pacific Ocean.

312 FAST FACT...

📖 **THE EARTH POLICY** Institute has said that it would take 1.5 Earths to sustain our present level of consumption.

313 GLOBALIZATION

🎓 **SINCE HUMANS FIRST FORMED SETTLEMENTS** and cultures, there has been interaction and exchanges of culture, ideas, and information as well as trade.

Globalization consists of four main areas: migration and movement of people, trade and transactions, capital and investment and dissemination of knowledge. Each of these can affect populations, such as increasing numbers of people migrating from rural to urban areas in search of jobs provided by transnational companies.

314 FAST FACT...

📖 **A GROWING ECONOMY** is often linked to a fall in crude death rates as medical resources, nutrition and sanitation improve, leading to population growth.

Recently, migration has been affected by the globalization of terror, which has tightened border controls and reduced immigration in countries like the US and UK.

Education is also key in managing population growth as people can make informed decisions about birth control, something that has been made easier through the availability of information and technology. The spread of technology has also meant a decreasing population density in some areas, as telecommuting becomes possible for high tech companies and service industries, which can operate and communicate online.

315 FAST FACT...

📖 **AS OF 2012**, an estimated 2.4 billion people were online, increasing the dissemination of knowledge and broadening our world view.

316 CULTURAL ASSIMILATION

🎓 **MIGRATION HAS PLAYED A HUGE ROLE** in the history of human development, and as different societies merge, cultural assimilation – when a group's native language and culture is lost to a dominant group – is what happens.

This can occur both in colonial areas where a dominant state has expanded into new territories as well as when mass immigration leads to a fundamental change in a society. There have been many examples of cultural assimilation, such as the adoption of Latin in Europe following the conquest of the Roman Empire. In the Americas, the conquistadores culturally assimilated the Aztec and Zapotec tribes in the 15th, 16th and, 17th centuries.

There are four key areas of immigrant assimilation; socioeconomic status, geographic distribution, second language attainment, and intermarriage. Full assimilation occurs when older members and newer members of society become indistinguishable.

317 FAST FACT...

📖 **BETWEEN 1880 AND 1920**, the United States took in approximately 24 million immigrants.

318 FAST FACT...

📖 **IMMIGRANT NAME** changing is also viewed as key indicator of cultural assimilation by sociologists.

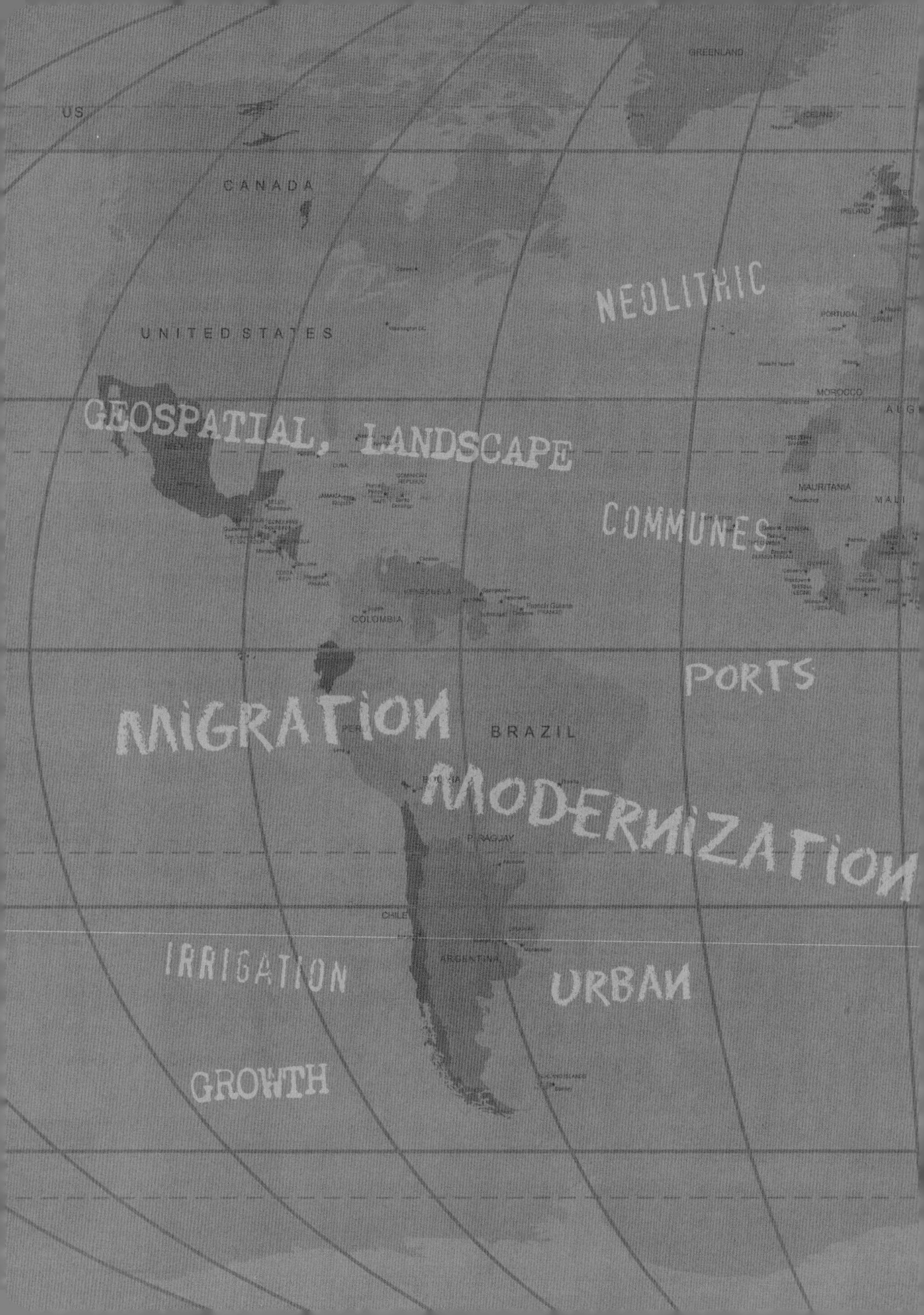
NEOLITHIC
GEOSPATIAL, LANDSCAPE
COMMUNES
PORTS
MIGRATION
MODERNIZATION
IRRIGATION
URBAN
GROWTH
US
CANADA
GREENLAND
UNITED STATES
PORTUGAL
MOROCCO
MAURITANIA
MALI
COLOMBIA
BRAZIL
PARAGUAY
CHILE
ARGENTINA

Settlement

501
DECAY
MEDCS
LEDCS
RURAL
RUSSIA
CHINA
IRAN
AFGHANISTAN
PAKISTAN
INDIA
LIBYA
EGYPT
CHAD
SUDAN
ETHIOPIA
SOMALIA
MALDIVES
SRI LANKA
BURMA
LAOS
THAILAND
PHILIPPINES
JAPAN
INDONESIA
DEMOCRATIC REPUBLIC OF THE CONGO
TANZANIA
MOZAMBIQUE
MADAGASCAR
NAMIBIA
BOTSWANA
SOUTH AFRICA
AUSTRALI
Moscow

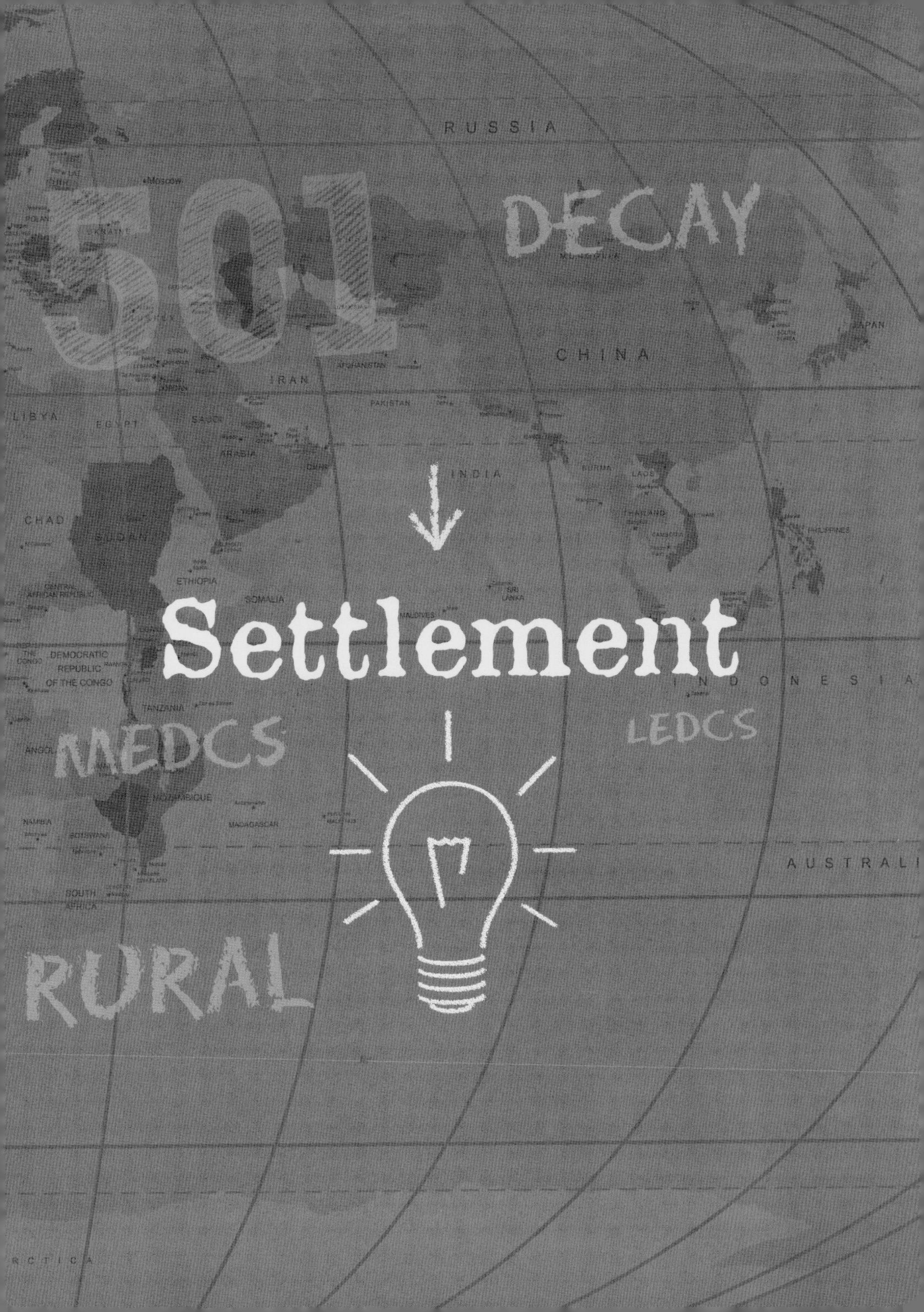

319 NEOLITHIC REVOLUTION

🎓 **THE NEOLITHIC REVOLUTION**, also known as the Neolithic Demographic Revolution or Agricultural Revolution, describes the period of history when humans established the first settlements.

According to carbon dating, the Neolithic revolution began about 12,000 years ago and developed over a 5,000-year period. It saw the transition of groups of humans from hunter-gatherers into societies that used crop rotation and irrigation methods to build up food supplies. Animals were domesticated and farmed for meat and produce.

320 FAST FACT...

📖 **THE FIRST** high density populations were quite a lot smaller than today's cities - Jericho, founded around 11-12,000 years ago had a population of 1-2,000.

This is turn led to the creation of fixed villages and towns where high density populations could be supported, and goods and services exchanged or bartered. The development of human culture including non portable art and architecture, political structures and writing became possible, with the first full Neolithic complexes appearing in Middle Eastern Sumerian cities about 5,500 years ago.

There are many theories on the factors that drove humans to take up agriculture including climate changes, power displays and evolutionary adaptation of plants and humans.

321   SETTLEMENT HIERARCHY

🎓 **SETTLEMENT HIERARCHY** is the ordering of settlements based on specific criteria such as population. A commonly used hierarchy would be capital city, regional center, large town, small town, village, and hamlet, with the greatest population, land area and number of services at the top

A key factor in determining the hierarchy of a settlement will be the services it provides, as settlements depend on each other and serve areas larger than the settlement boundaries, known as the settlement's sphere of influence. Smaller settlements like villages offer low order goods and services, such as daily groceries, and have a smaller sphere of influence. Larger towns and cities provide high order goods and services that people will travel to access, increasing the city's sphere of influence and moving it further up the settlement hierarchy.

Neighborhood shopping centers consist of shops selling inexpensive, daily items, which provides for an area within a town and have been impacted by the rise of out of town shopping centers and malls in recent years.

322   FAST FACT...

📖 **MANILA** in the Philippines has the greatest population density of any city in the world with over 1.65 million people in 14.9 square miles.

323   FAST FACT...

📖 **THE ISLAND** of Tristan de Cunha is the most physically isolated location on Earth. The nearest neighbor to its 270 residents is South Africa, around 1750 miles east.

324 CENTRAL PLACE THEORY

PROPOSED BY GERMAN GEOGRAPHER Walter Christaller in 1933, Central Place Theory is a theory that tries to explain the reasons why settlements are located where they are.

He used the assumption that the land in areas he was studying was flat, to remove the variable of barriers, and that the population was evenly distributed. He also assumed that people will always purchase goods from the nearest available place, and that whenever demand for a product or service increases it will be offered in close proximity to the population.

Typical Central Place Theory models have major cities at the three points of a triangle superimposed on an area, and three additional cities in between those points to make a classic hexagon, as this is the most efficient shape to serve areas without any overlap.

326 FAST FACT...

RURAL AREAS such as Iowa and Wisconsin in the US are two areas that have come closest to matching CPT.

CPT has three principles: marketing, transportation and administration based around the constants three, four and seven, so each market central place is three times bigger than the next market area down the hierarchy, each transportation central place is four times larger than the next lowest and each ̶ ̶ ̶ven times larger than the next lowest.

325 FAST FACT...

IN THE 1950s, Central Place Theory was used to restructure boundaries in Germany and the system remains in place to this day.

327 CITIES

A CITY IS A LARGE, permanent settlement that may have a particular legal, historical or administrative status according to local custom, although there is no international agreement on what defines a city.

Cities commonly have a controlled infrastructure to maintain sanitation, housing, transportation, utilities and land use, with larger cities often having suburban residential areas surrounding one or more central business districts.

Traditionally, cities were founded in areas with abundant natural resources such as rivers, oceans or seas and forests that would provide a supply of food and building materials. Modernization and development of transportation has meant cities can be established in remote areas such as Las Vegas in the Mojave Desert.

Several claims are made for the title of world's oldest city including Damascus in Syria, which has evidence suggesting the site was occupied over 8,000 years ago.

328 FAST FACT...

THE HIGHEST city in the world is considered to be La Rinconada in Peru. Located near a gold mine, the city lies at 16,700 feet above sea level.

329 FAST FACT...

THE VATICAN CITY has the highest per capita crime rate in the world, with just over 800 residents and 600 crimes committed per year.

330 URBAN GROWTH

URBAN GROWTH is the rate at which the human population of urban areas expands. Not to be confused with urbanization, which is the change in proportion of a population living in urban areas compared to rural ones, Urban Growth is represented as a number not a percentage or ratio.

Currently, the world's urban population is growing at four times the rate of rural areas with the urban population projected to double to more than five billion between 1990 and 2025.

Attempts to limit the physical size of urban development are known as an Urban Growth Boundaries, which can have dramatic affects on the price of housing.

Organization such as the United Nations have pointed to the positives of urban growth as drivers of development, enhancing employment, creating wealth, innovation, and knowledge.

331 FAST FACT...

OVER THE 20TH CENTURY the urban population jumped from 220 million in 1900 to 732 million in 1950 and 3.2 billion in 2005.

332 FAST FACT...

MOTIVES THAT CAUSE people to leave the countryside such as lack of jobs are called push factors, incentives such as better health care causing people to go to cities are called pull factors.

333 URBAN DECAY

👆 **URBAN DECAY IS THE WAY** in which part or all of a working city falls into disrepair and disuse and ultimately can lead to the creation of ghost towns.

The reasons for urban decay can include a major change in the economy, transportation routes, industry or natural resources, as is the case for many ghost towns in the US.

During the 19th and 20th centuries, many small towns were established based on natural resources that funded them. As the resources dried up, the towns became financially unviable and the population migrated. More recently, changes in industry have lead to urban decay in places like Detroit as vehicle manufacturing has relocated overseas and recession has impacted housing developments in Ireland, Spain, and China leaving many residential areas empty.

334 FAST FACT...

📖 **IN CHENGGONG**, Yunnan, in China, there are more than 100,000 apartments with no residents. It is considered Asia's largest ghost town.

335 FAST FACT...

📖 **PYRAMIDEN**, a former Soviet mining town in Norway, was abandoned in 1998. Due to the freezing climate, the buildings will still be visible 500 years from now.

Sometimes environmental disasters can lead to urban decay, most famously in Chernobyl. From 1986 to 2000, over 350,000 people were evacuated from the area following a nuclear meltdown.

🎓 **URBAN LAND USE IS A WAY** of describing the way land is used in a town or city based on categories such as commercial, residential, entertainment, open space, industrial, and services.

These are split into five different zones for urban areas; the Central Business District with the majority of shops and public buildings, the Inner City manufacturing area, the Inner City residential area, the Inner Suburbs, and Outer Suburbs. The first inner city area, zone two, is also called the twilight or transitional zone.

The CBD is in great demand, pushing up prices as well as the height of buildings, as it is accessible via public transport and is usually the oldest part of a city. On the outskirts of a city is the rural-urban fringe, which can cause friction between different land uses such as industry looking for cheap land and residents who want an attractive environment.

337   FAST FACT...

📖 **IN 2013 BUILDING** work started on the Kingdom Tower in Jeddah, Saudi Arabia – when completed the tower will stand over half a mile (3,280 feet) tall.

338   FAST FACT...

📖 **LONDON HAS** the oldest subway or metro system in the world. Known as the London Underground, it was first opened in 1863.

339 URBAN LAND USE – LEDCs

🎓 **LESS ECONOMICALLY DEVELOPED COUNTRIES** have similar land use needs to more economically developed countries but the pattern of land use in urban areas is different. Although each city has its own characteristics, a typical LEDC city has a Central Business District (CBD), which is often the oldest part of the city.

In contrast to MEDC cities whose suburban fringe is very often a place of high quality housing, in LEDCs, the poorest housing, known as squatter settlements or shanty towns, is found on the edge of the city. High-class housing can be found around the edge of the CBD and in a spine radiating out towards the edge of the urban area. This is likely to be a main transport route and a desirable street, possibly a former colonial area where houses or apartments will have space for servants. People living here will need easy access to the CBD.

340 FAST FACT...

📖 **THERE CAN BE A STARK** contrast between high rise buildings of a CBD in a LEDC city and the squatter settlements found on the suburban fringe.

341 LOST CITIES

🎓 SOMETIMES, DUE TO A CHANGE in climate or other natural disaster, social upheaval or war, a formerly populated area can become abandoned and its location forgotten. These places are known as Lost Cities.

There are two broad categories for a Lost City, either its existence is completely unknown until its rediscovery, or it is known of but its precise location isn't certain and only exists in rumor, myths or historical records.

Cities can be lost due to natural resources such as fresh water sources drying up, or from a natural disaster such as the earthquake that caused the ancient Greek city of He like to become submerged in the Corinthian Gulf. Sometimes humans are directly responsible, as with one of the most famous lost cities, Machu Picchu, where the population was eradicated by smallpox and the city forgotten.

342 FAST FACT...

📖 **NUMEROUS LOST CITIES** are reported in South America, often in remote parts of the Amazon. Although probably mythical, some have been discovered only recently.

343 FAST FACT...

📖 **ADVANCES IN TECHNOLOGY** have enabled many lost cities to be rediscovered using satellite imagery and aerial photography.

344 RURAL

 **RURAL AREAS ARE THE PARTS** of a country not covered by urban areas such as towns and cities, including all populations, housing, and territory. They can generally be divided into accessible rural and remote rural areas.

Compared to urban areas, rural areas are less densely populated and have fewer transport networks. Economies are based on tourism, farming profitability and jobs that use natural resources. In many countries, rural areas are under pressure from forest clearance, mining and cash crops such as sugar cane, which is used for biofuels. In countries such as Brazil there is considerable conflict over the use of rural areas for cash crops that require deforestation.

Many wealthy urban residents often buy second homes or retirement properties in rural areas and national parks. This pushes up housing prices in in-demand areas as well as reducing the number of houses available for local residents.

345 FAST FACT...

📖 **OVER THE NEXT 20 YEARS**, 350 million people will move from rural to urban areas in China.

346 AGRICULTURE

BEFORE INDUSTRIALIZATION, the population of a country was largely employed in agriculture to ensure food production. Because of this, many settlements were located in areas that were best suited to agricultural purposes.

The factors that make a good location for a farming settlement are a combination of physical factors, such as altitude, temperature and rainfall, as well as human factors that can have an effect at a national or regional level.

Some of the key human factors are transport, especially for products that are bulky, such as potatoes, or do not stay fresh for long, a market to ensure demand profitability, technology to ensure the efficiency of production and government policies that may affect imports and exports, which will have a knock-on effect quality of life for farmers.

347 FAST FACT...

ISRAEL was primarily desert until desalination plants were constructed to create a source of water.

348 FAST FACT...

IN THE US, the face of agriculture changed when in the late 1800s the American railroad system became a nationwide transportation network.

349 IRRIGATION

IRRIGATION IS THE ACT OF APPLYING WATER TO LAND to assist in growing plants or crops as well as maintaining landscapes, suppressing dust, disposing of sewage, and in mining.

The need for water is one of the key factors in any human settlement and has lead to the greatest population densities forming close to water sources such as rivers. One of the most famous examples of this is in Egypt, where 95% of the population live within 12 miles/20 kilometers of the River Nile.

Ancient Egyptians utilized the flooding of the Nile to extend the areas where they could grow crops by digging dykes around land plots. Evidence of terrace irrigation has been found in Aztec civilizations dating back 6,000 years.

By the 20th century, diesel and electric motors were used to pump groundwater our of aquifers and around 689 million acres/1.08 million sq miles (2.8 million sq kms) of land world wide had irrigation infrastructure.

350 FAST FACT...

MORE AFRICANS have access to mobile phones than clean drinking water.

351 FAST FACT...

THE IRRIGATED land area world wide occupies about 16% of the total agricultural area, but supplies about 40% of crop yield.

352 PORTS

THROUGHOUT THE HISTORY of human settlements, ports have played key roles in the development of a country as places with at least one harbor where ships can dock to transfer people or cargo.

Located on coasts or shorelines, ports can take advantage of natural harbors such as Kingston Harbour in Jamaica, or they can be based around artificial harbors like Rotterdam in the Netherlands.

Ports can vary a great deal in size and purpose, from small fishing ports based around recreation through to massive international cargo ports that extend for miles. Sometimes ports can become obsolete due to coastal erosion, or cease to become economical due to the increasing size of ships, as happened with London.

Ports located on rivers are known as river ports and generally deal with shallow draft vessels such as barges.

353 FAST FACT...

THE TITLE OF 'BUSIEST PORT' is contested, but currently the record for busiest container port is Shanghai, handling over 30 million TEU cargo containers in 2011.

354 FAST FACT...

PORTS ARE under increasing scrutiny to decrease their environmental effect as pollution can damage the local ecosystem.

355 ADMINISTRATIVE DIVISION

WITHIN THE BORDERS OF A COUNTRY, there are many other regions containing groups of settlements and under separate local government jurisdiction. These are known as Administrative Divisions.

Administrative Divisions include a wide spectrum of sizes, from countries like Scotland being constituent parts of the UK, countries in turn may be divided into counties and then into municipalities. There are a wide range of terms for each sub-division that vary from nation to nation, such as arrondissement or barrio, equating to districts in French and Spanish speaking countries.

The purpose of Administrative Divisions is to provide citizens of a specific area with some degree of autonomy from state rule via local government. In the US for example, states were able to vote on the legalization of gay marriage, providing an opportunity for Administrative Divisions to decide on laws based on the attitudes of the local population instead of national law.

356 FAST FACT...

NATIVE AMERICAN tribes manage American Indian reservations so laws may vary from federal law and the tribe may have jurisdiction not federal government.

COMMUNES

ALTHOUGH THE TRADITIONAL IDEA of a commune is often one of long-haired students and psychedelic music, a commune by definition is an intentional community of people who live together and share common resources, values, and frequently work and income.

The main principle of a commune is the idea of putting the group before the individual and replacing the traditional hierarchy of a settlement with a more egalitarian format where everyone is equal.

Although they are strongly tied to communist and socialist political theory, one of the longest and most successful communes in the US was founded by the Shakers religious organization. They established a commune in New York in 1776 and their communities still exist today.

Communes exist worldwide, notably in Israel where the Kibbutz is an official commune, as well as Germany and Venezuela.

358 **FAST FACT...**

IN ISRAEL in 2010 there were a total of 270 Kibbutzim that accounted for 40% of the agricultural output.

359 **FAST FACT...**

VENEZUELA has initiated over 200 socialist communes that are independent of the government.

360 MIGRATION

MIGRATION IS THE MOVEMENT of a population from one area to another, sometimes over a long period of time and involving large groups. Migration can be voluntary as is the case with nomadic groups throughout history, or involuntary as with people who have been displaced by war or famine.

Humans have a history of being migrants that started with the movement of humans from Africa across Eurasia around one million years ago. Homo Sapiens had spread across Australia, Asia, and Europe around 40,000 years ago, while migration to the Americas took place 15-20,000 years ago.

Migration can also be internal when people move within the same country or region, or international when people migrate from one country to another. The reasons for this include economic – for work or a career, social – to find a better quality of life to be with family or friends, political – to escape persecution, or environmental to escape natural disasters.

361 FAST FACT...

EMIGRATION is when someone leaves a country; immigration is when someone enters a country.

362 FAST FACT...

A REFUGEE is someone who has left their home but doesn't have a new home to go to.

363 LANDSCAPE HISTORY

OVER THOUSANDS OF YEARS, humans have had a significant impact on the physical appearance of the planet, the study of this is called landscape history.

Landscape history incorporates a wide range of human effects on the land, from specific individual features to hundreds of square miles. Topics that are studied include settlement morphology – the way settlements change and whether they are dispersed or nucleated, field systems and boundaries which can indicate previous layouts to settlements, place names used to signify prominent features of the landscape at that time, and the status of the settlement compared to others nearby.

Studies in Landscape History combine research of historical documents and aerial photography as well as local historical knowledge with fieldwork such as physical excavation to identify earthworks – an archeological term for artificial changes in land level such as hill forts.

364 FAST FACT...

SPECIES OF PLANTS and trees are sometimes known indicator species because they can identify previous land use, for example bluebells suggest ancient woodland.

365 GEOSPATIAL MODELING

WHEN IT COMES to predicting trends and events in the future, analyzing geographical elements has a significant contribution using something called Geospatial Modeling.

This works on the concept that events taking place, such as terrorist attacks, are neither random nor uniform in their occurrence, but predisposed towards certain areas.

Geospatial Modeling presents a simplified representation of a geographic reality, allowing a focus on key variables or events. There are two main types of Geospatial Modeling: Inductive and Deductive.

Inductive methods use the spatial relationship between known events and environmental factors such as topography and infrastructure. Often this is displayed as a map with graphic overlays.

Deductive modeling requires expert knowledge to describe the relationship between a known event and environmental criteria such as the elevation of the land or proximity to water. Because of the human element required, this can be subjective.

366 FAST FACT...

GEOSPATIAL MODELING can be used to create 'crime maps' showing the likelihood of specific crimes occurring in a certain area.

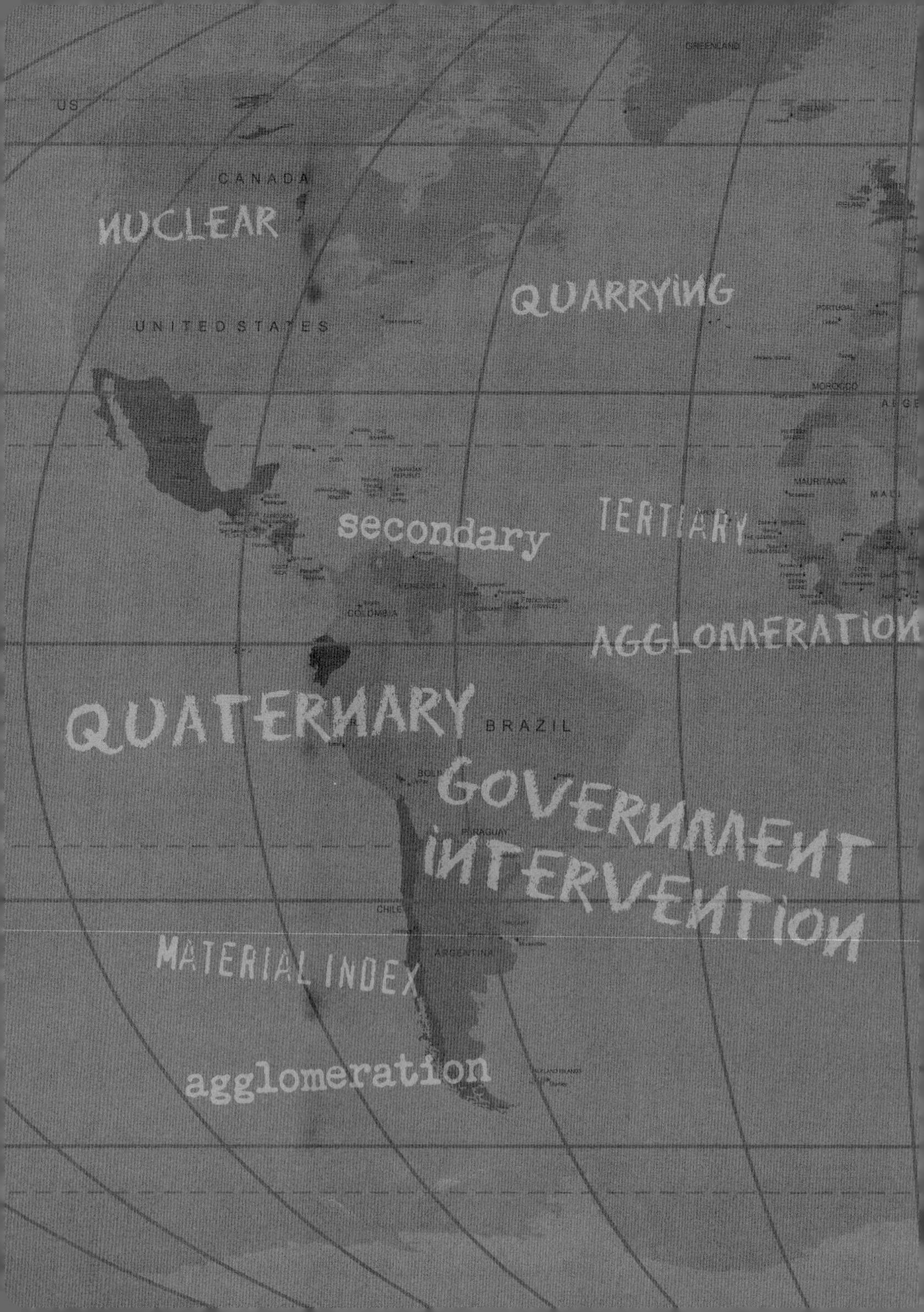

NUCLEAR

QUARRYING

secondary

TERTIARY

AGGLOMERATION

QUATERNARY

GOVERNMENT
INTERVENTION

MATERIAL INDEX

agglomeration

FOOTLOOSE
RUSSIA
501
informal
CARBON CREDITS
DESERTIFICATION

Industry & Energy

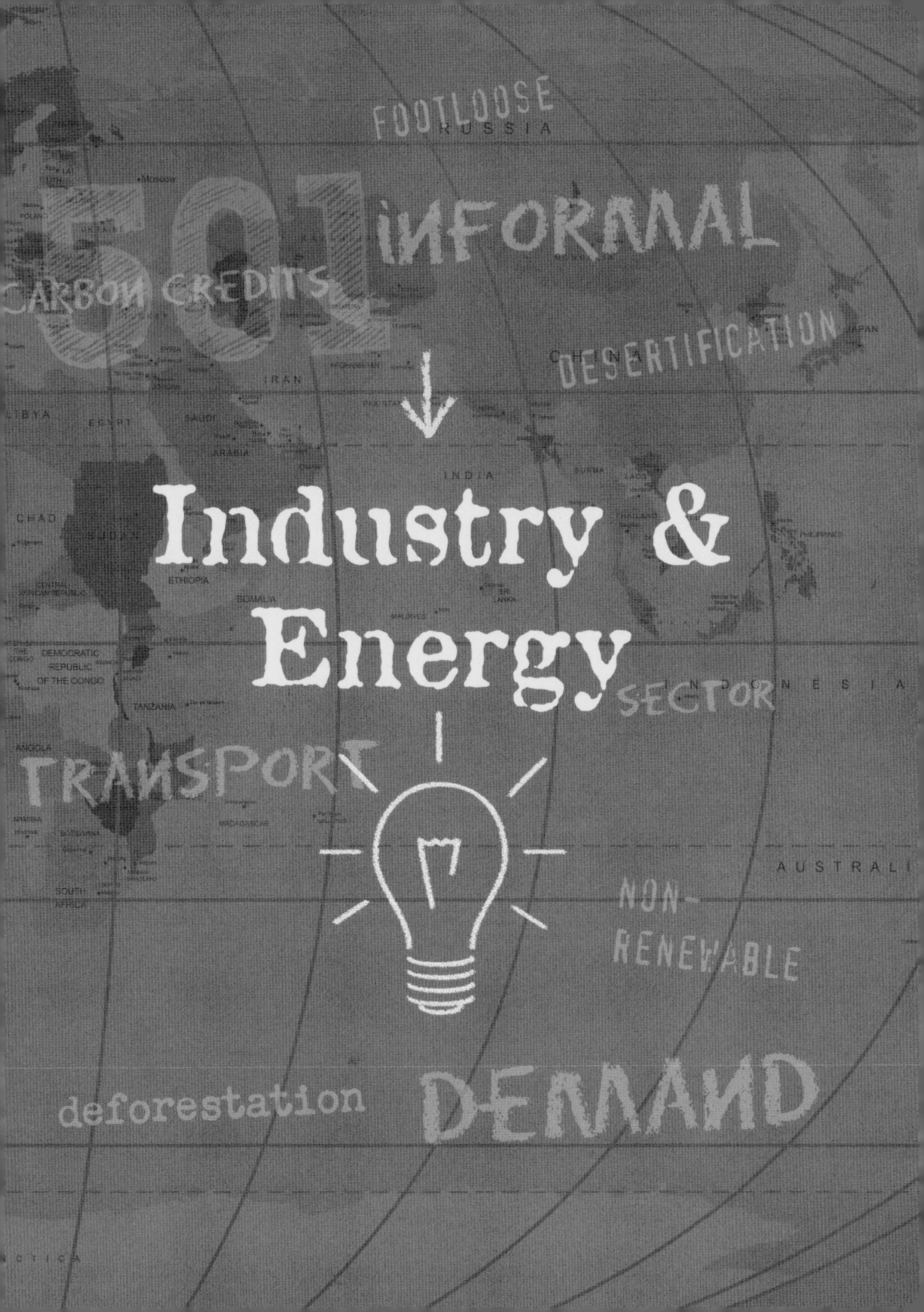

TRANSPORT
SECTOR
INDONESIA
AUSTRALI
NON-
RENEWABLE
deforestation DEMAND

367 TYPES OF INDUSTRY

🎓 **INDUSTRY IS CLASSIFIED** into different sectors – primary, secondary, tertiary, and quaternary. The employment structure of a country shows how the labor force is divided into the different sectors.

Primary – involves extracting resources from the land or sea and includes farming, fishing, forestry, coal mining, oil drilling, etc. A primary industry is located at the source of the raw material.

Secondary – where products are made by processing raw materials or assembling components. Employment in secondary industries is in decline in the UK and USA because cheaper labor can be found in SE Asia.

368 FAST FACT...

📖 **INDUSTRY CAN BE SEEN** as a system with inputs such as raw materials, processes like making or assembling products, and outputs where finished products are on sale to consumers.

369 FAST FACT...

📖 **IN THE RICHEST** country (USA), most people work in the tertiary sector & in the poorest country (Nepal), most people work in the primary sector.

Tertiary – those industries that provide a service, e.g., health, administration, retailing, and transport.

Quaternary – those industries providing information services, such as computing, ICT (information and communication technologies), consultancy (offering advice to businesses) and R&D (research, particularly in scientific fields).

INDUSTRIAL LOCATION

THERE ARE MANY FACTORS affecting the location of industry and these have changed over time. Different industries require different inputs of course, and traditionally, have been most likely to locate where these inputs are readily and cheaply available. On the other hand, some industries are described as 'weight-gaining', meaning that they gain weight or bulk during manufacture, for example, baking, brewing, and soft drink manufacture. These weight gaining industries used to be located near to their markets in order to reduce transport costs. As transport has become more efficient though, the influence of the raw materials on an industry's location has been reduced.

The distribution of power has also changed the way industries locate. Where industries used to rely on steam power and needed to be located near to coalfields for example, today they use electricity, which is generated in power stations and transmitted over long distances.

Lower operating costs and labor supply, as well as fewer government regulations also play a huge part in the location of industries.

371 FAST FACT...

FACTORIES BASED in countries such as India and China have around 25% lower operating costs than the UK.

372 FAST FACT...

MANY CALL CENTERS for insurance companies, banks, and railways are now located in countries which have cheap but skilled labor.

373 GOVERNMENT INTERVENTION – AFFECTING INDUSTRIAL LOCATION

GOVERNMENTS play a significant role when it comes to influencing industrial location. Examples of why they do this include:

- to move industrial activity out of congested urban areas into less prosperous areas, e g., Mumbai and Kolkata in India.
- to move industry into derelict brownfield sites in the inner parts of urban areas, e g., the urban development corporations set up in London Docklands.
- to create jobs by setting aside plots of land for industrial estates with good road, air, sea or rail access and other services.
- to improve air quality by banning industries that create pollution from locating near to residential areas.
- to develop an area by offering loans, subsidies, and tax exemptions to companies.
- to attract industry by providing stable government without corruption.

374 FAST FACT...

GOVERNMENTS in developed countries have attracted science, business, and retail parks to parkland with lakes on greenfield sites on the edges of urban areas.

375 FAST FACT...

'GROWTH POLES' have been established in rural Brazil to attract people away from the crowded areas around Rio de Janeiro and Sao Paolo.

376 WEBER'S MATERIAL INDEX

ALFRED WEBER WAS A GERMAN spatial economist who, in 1909, devised a model to try to explain and predict the location of industry. This model is renowned in the history of geography although it no longer relates to modern conditions including different stages of economic development.

Weber's model assumed that industrialists choose a least-cost location for the development of new industry. The theory is based on a number of assumptions, among them that markets are fixed at certain specific points, that transport costs are proportional to the weight of the goods and the distance covered by a raw material or a finished product, that perfect competition exists, and that decisions are made on a rational economic basis.

377 FAST FACT...

WEBER'S MODEL was challenged by an alternative 'area of maximum profit' approach put forward by David Smith in 1971.

Weber postulated that raw materials and markets would exert a 'pull' on the location of an industry through transport costs. Industries with a high material index would be pulled towards the raw material and industries with a low material index would be pulled towards the market.

AGGLOMERATION AND FOOTLOOSE INDUSTRIES

THESE ARE TWO 'SPECIAL CASES' of industrial location; agglomeration and footloose industries.

Agglomeration is when a number of producers in the same or related industries group themselves together. They do this to benefit from local skill pools, economies of scale or the prowess of a locality in a particular field. Examples include the large number of financial services companies (e.g., banks and insurance companies) which are headquartered in the City of London, or the large technology corporations as well as thousands of small start-ups which are located in Silicon Valley, California.

Footloose industries are those that are less dependent on factors that tie them to a specific geographical location. Unlike manufacturing industries, tertiary or services, companies do not have to be near a source of raw materials. As long as they have suitable transport, energy and communications links, they can locate themselves virtually anywhere in the world. Examples of footloose industries are computer software development, telephone sales, and call centers.

379 FAST FACT...

THE RISE OF HIGH technology industries in the UK has meant areas like the M4 Corridor, with many footloose industries like Hewlett Packard & Sony that attract skilled workers and have excellent communication links have benefited from growth.

380 INFORMAL SECTOR

🎓 **IN DEVELOPING COUNTRIES,** a large and growing number of people have found or created their own jobs in the informal sector. This covers a wide variety of activities meeting local demands for a wide range of goods and services. The informal sector incorporates sole proprietors, cottage industries, self-employed artisans, and even moonlighters (someone who holds a second job after hours). They are manufactures, traders, transporters, builders, tailors, shoe makers, mechanics, electricians, plumbers, etc.

Some governments recognize the importance of these small scale enterprises, which apart from creating employment, provide goods at affordable prices. And, a few commercial banks extend loans to entrepreneurs who are themselves forming co-operatives. India, under the 'Small Industries Development Organisation' is one country which actively encourages the growth of co-operatives by setting up district offices that offer technical and financial advice for the informal sector.

381 FAST FACT...

📖 **CHILDREN,** many of whom may be under the age of 10, form a significant proportion of informal-sector workers.

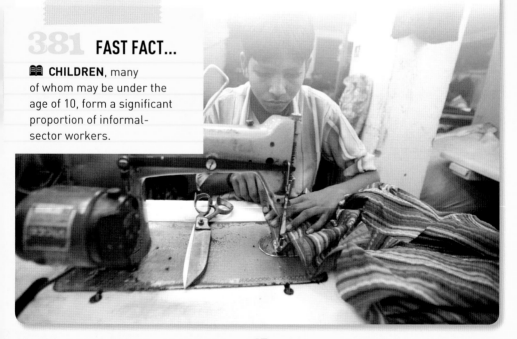

PRIOR TO ABOUT 1890, foot, carriage or horse-drawn trolley accomplished most movement of people and goods but modern transportation has completely changed the way that industry operates in the world. Today, of course it's possible to transport goods from one side of the world to the other within just hours or days and the development of a new generation of airships are set to revolutionize, even further, the way cargo is transported.

Not just international air travel and shipping but transportation methods within one country are vitally important and can even foster the growth of whole cities. Within the US for example, where pioneers of business and exploration first used the Mississippi river system in the 1800s, today use it to carry agricultural and manufactured goods, iron, steel, and mine products from one end of the country to the other. The Mississippi River and the Missouri River, the two major stretches of the system, see 423.2 million metric tons of freight transported every year. This has nurtured the growth of major cities including Minneapolis, St. Louis, Memphis and New Orleans.

383   FAST FACT...

AROUND 50,000 merchant ships cruise the world's oceans and seas transporting goods.

384   FAST FACT...

THE NEED FOR A DRAMATIC reduction in carbon emissions will challenge global transport networks in the future.

385 GLOBALIZATION OF INDUSTRY

GLOBALIZATION OF INDUSTRY has meant that more MEDCs and LEDCs are working together. Often the management skills and finance in MEDCs are combined with the plentiful, cheap, and hardworking labor in LEDCs with the operations of many transnational companies based on this principle.

Meanwhile, the decline of industries in some MEDCs has created economic and many associated social problems. The decline of heavy industry in the Ruhr in Germany for example, demonstrates that globalization doesn't always work in the favor of MEDCs. Many of the original raw materials in the Ruhr are exhausted, and there are high labor costs and old, outdated machinery, which has led to cheaper steel being imported from south east Asia where labor costs are lower. Whilst the Ruhr has been in a state of degeneration it's also a good example of regeneration as new industries are moving in and, as industries have changed, so the workforce have learnt new skills.

386 FAST FACT...

📖 **IT'S THOUGHT THAT TRANSNATIONAL** industries are more competitive than the rest, since their exposure forces them to be and their position allows them to first see opportunities as they arise in world markets.

387 FAST FACT...

📖 **THE FORD MOTOR COMPANY** is a TNC with headquarters in Detroit, USA and factories producing parts in at least 20 different countries.

388 ECONOMIC ACTIVITY –CONSEQUENCES

THE CONSEQUENCES OF ECONOMIC ACTIVITY are such that not only are ecosystems being destroyed and glaciated areas being exploited, but also long-term sustainability is threatened.

Rainforests (especially the Brazilian rainforest) are being destroyed due to deforestation and in order to gain income, and the rise in the building of tourist attractions including ski resorts in glaciated areas, pose a threat to the fragile environment (e.g., soil erosion can interfere with flora and fauna).

Meanwhile, illegal logging (in violation of national laws in Indonesia and Peru, for example), poaching and illegal fishing, use of fossil fuels and building of new roads, motorways and airport runways that encourage travel are all activities that pose a threat to sustainability. For the long term future, sustainable management of these areas needs to take place.

389 FAST FACT...

AS COUNTRIES DEVELOP and demand for energy and resources continues to grow, international co-operation is paramount to a sustainable future.

390 RESOURCES AND ENERGY – GROWING DEMAND

🎓 **THE DEMAND FOR RESOURCES AND ENERGY** grows as population increases and economic development spreads to more countries. The demand for resources and energy isn't evenly spread across the world though. Whilst MEDCs have only about 25% of the population, they consume about 80% of the energy produced. What's more, about 2 billion people in the world are still without access to modern energy.

As resources and energy is being exploited, there are many socioeconomic, environmental and political consequences.

- As demand increases and resources are depleted, prices rise causing the cost of living to go up and subsequent loss of support for governments.
- Increased carbon emissions causes global warming with consequences including climate change.
- Air pollution is more prevalent.
- Ecosystems such as rainforest are under threat from exploitation as countries (such as Brazil) exploit their resources for development.
- There is an increased need for global agreements and international co-operation.

391 FAST FACT...

📖 **THE INTERNATIONAL** Energy Agency estimates that by 2030 the world energy demand will increase by 45%.

392 FAST FACT...

📖 **THE MAIN FORM OF ENERGY** today is electricity which can be obtained from fuels like oil and coal or in moving water and wind, for example.

393 NON-RENEWABLE RESOURCES

NON-RENEWABLE RESOURCES EXIST in finite quantities, i.e., once they've been used up they're gone and can't be renewed. Perhaps the most important and certainly worrying fact, is that many developed countries are overwhelmingly dependent on non-renewable fuel sources such as petroleum, coal and natural gas.

The management of these is therefore vital as is the investment in renewables.

Oil/Petroleum – major oil fields are found in the Middle East, Alaska, Iran, Iraq, and Russia. The oil industry is run by major global companies (e.g., Mobil, BP, and Shell) who explore, extract, refine and transport and market petroleum products. Next to air and water, petroleum is perhaps the most essential resource for developed countries. Globally, petroleum is being consumed at a rate of 3.1 billion barrels per year.

394 FAST FACT...

CURRENTLY, ABOUT 95% of the world's energy supply comes from non-renewable sources.

Coal – the world's coal supply greatly surpasses petroleum in quantity and longevity. It used to be the main fuel for industry, especially for iron and steel making and to heat homes. Today it's mainly used in power stations to produce electricity and contributes about 25% of the world's energy supplies.

Natural Gas – exists in abundance, and as far as fossil fuels go, has attractive qualities. It burns very cleanly (with little air pollution), requires little or no processing before use and can be transported cheaply overland by pipe. It isn't well suited to transoceanic trade though and supplies are limited.

395 FAST FACT...

THE WORLD'S FIRST commercial oil was drilled in Pennsylvania in 1859.

396 FAST FACT...

ALTHOUGH THERE ARE MANY COAL reserves left in the world, the cost of mining makes it uneconomic for many areas.

397 NUCLEAR POWER

SEEN AS A MAJOR source of energy for the future, nuclear energy is stored in uranium atoms and is released as heat, which turns water into steam that drives turbines thus creating electricity. Energy in a nuclear station is created when neutrons strike uranium 235. There are many advantages; reserves will last for many years and technology has the ability to create almost unlimited cheap power with relatively low environmental risks, refined uranium is relatively easy to transport and power stations don't need to be located near to raw materials.

Views are divided when it comes to nuclear power though. Those against nuclear power point to the accidents, such as the 1986 Chernobyl disaster in Russia, the dangers of military use, its inflexibility to meet peak demand and the problems of nuclear waste disposal including how to safely decommission a power station at the end of its life.

398 FAST FACT...

GLOBALLY, THERE ARE about 500 nuclear reactors, which produce nearly 20% of the world's energy supply.

399 FAST FACT...

CHINA AND INDIA as well as some LEDCs are now developing nuclear energy sites.

400 QUARRYING

FIRST USED BY EARLY settlers for building stone and extracting metal for weapons, quarrying continues today in many countries, as a primary industry that involves the extraction of rocks such as limestone and slate. Usually a type of open-pit mine, quarrying may also take place underground. Whilst there are obvious benefits – including demand for products, creation of jobs, income through taxation, as well as the creation of tourist attractions where flooded quarries have become lakes, for example - there are problems associated with quarrying.

Valuable agricultural land and wildlife habitats are destroyed. Quarries also create much visual, environmental and noise pollution, and to help control and eliminate the pollution of public roads by trucks as they leave a quarry, wheel washing systems are becoming more common.

Quarries in level areas with shallow groundwater or which are located close to surface water often have engineering problems with drainage. Generally the water is removed by pumping while the quarry is operational, but for high inflows more complex approaches may be required.

401 FAST FACT...

ONE OF THE MORE EFFECTIVE and famous examples of successful quarry restoration is Butchart Gardens in Victoria, BC, Canada.

402 WASTE AND POLLUTION

AS COUNTRIES DEVELOP and consumption increases so does the amount of waste per capita, and pollution becomes a greater problem.

The amount and type of waste produced varies between countries. Due to higher levels of consumption, MEDCs produce more waste, but in LEDCs less packaging in used on products, disposable items are used less, lower literacy levels means there is less production of written material and it's more common to reuse items therefore creating less waste.

Pollution (air, water, and noise) meanwhile, is largely caused by manufacturing industries, found mainly in developed countries, and can affect the environment and humans in a number of ways. It may interfere with natural processes like weather for example, it can impact on livelihoods (pollution of the sea may affect those involved in the fishing and tourism industries) and general wellbeing.

There are global, national, and local strategies in place to reduce levels of waste and pollution.

403 FAST FACT...

SOME MEDCs PRODUCE over 700 kgs of waste per person per year whilst in LEDCs the figure is around 330.7 pounds per person per year.

404 FAST FACT...

IN APRIL 2010, a deep water oil well exploded in the Gulf of Mexico, killing 11 people and injuring 17 others.

405 WATER USAGE

THE AMOUNT OF WATER used in the world everyday is very uneven. MEDCs use more water than LEDCs – households, farming, and industry all demand water.

In general LEDCs, like Bangladesh and Malawi for example, use most of their water in agriculture and little in industry or domestic use. Their irrigation channels are prone to loosing a lot of water through evaporation though. By contrast, in MEDCs irrigation is mechanized. That is, vast amounts of water can be released via sprinklers and timed irrigation systems, at the touch of a button.

MEDCs have a more significant use of water for domestic reasons, and tend also to have a higher percentage for industrial use. There are exceptions to this though. The USA for example, still has a high amount of water used for agriculture due to the volume of farming across the country.

The way the world uses water is changing though. As more multinational companies locate in LEDCs there will be more demand on water for industrial use and as MEDCs accrue greater wealth, there is more demand for showers, baths, washing machines, swimming pools and even spas and golf courses, which require vast quantities of water. In LEDCs though, many people don't have access to piped water – instead it's brought to the home from a well or stream – and so use it more sparingly.

406 FAST FACT...

THE AVERAGE PERSON in the developing world uses 2.6 Gallons (10 liters) of water per day. The average North American uses 105 Gallons (400 liters) per day.

407 FAST FACT...

IN INDIA, COCA-COLA uses over 264,172 gallons of water a day to produce drinks.

🎓 **THERE IS A HIGH DEMAND** for water in MEDCs but is this met by supply? The answer is; not always. In the UK for example, there's a big issue with water supply. Areas which receive high amounts of rainfall tend to be sparsely populated and whilst one – third of the population live in South East England, this is the driest area in the country. Ways to manage the water supply are therefore vital and include making sure broken pipes are mended, and using reservoirs and dams in one area to pipe water into large urban areas.

Domestic water use can also be monitored and it's known that households with water meters use less water than those without. Industries can also look to recycle waste water. For example, when using water for cooling in steel-making, the water can be recycled again and again in the process. In agriculture, drip-feed irrigation systems could be used rather than sprinkler systems.

409 FAST FACT...

📖 **IN DEVELOPED COUNTRIES,**
as much as 30% of water loss can occur through broken pipes.

410 MANAGING WATER SUPPLY – LESS DEVELOPED COUNTRIES

THERE ARE MANY PROBLEMS when it comes to supplying clean water in LEDCs and most of them are related to contamination, although sometimes of course, water is scarce in the first place and in some cases, salt water is contaminating groundwater making the problem worse.

Chemical fertilizers and industrial wastes seep into the soil and contaminate aquifers in developed countries but in less developed countries the lack of appropriate technologies are often the route of the problem. Wells, dug by hand, are a common way of accessing water but the supply can be unreliable and sometimes the well itself can be a source of disease. If the resources are available to dig a bore hole and either a hand or diesel pump is used, water can be brought safely to the surface.

In many cases, funding isn't available for safe water supply although government and community-led sanitation projects are in place, which include foreign investment as well as strategies, such as recycling waste water to use on crops, growing crops less dependent on a high water supply and improving irrigation techniques to help reduce the need for water.

411 FAST FACT...

60 MILLION CHILDREN are born each year in LEDCs who do not have access to safe water.

412 DEFORESTATION

DEFORESTATION, WHICH IS THE CUTTING DOWN and removal of areas of forest, has been increasing rapidly in the last 20 years. Extension of commercial agriculture, timber extraction for hardwood production, mining operations, road building and exploitative recreation and tourism opportunities are all reasons for deforestation. Forests are even destroyed to help develop electricity supplies. Hydroelectric power projects for example, need reliable high rainfall as is characteristic of rainforests.

Most rainforests of course are located in LEDCs, e.g., Brazil and Indonesia and in these cases, deforestation provides jobs in mining, logging, and tourism as well as income from the export of timber and ores, which increases the country's GDP. There are many problems though, most poignantly that soils are being left infertile, water supplies are polluted, and the indigenous population are losing their livelihoods, homes, and culture. Due to an increase in CO_2 in the atmosphere, deforestation has also been related to global warming.

413 FAST FACT...

50% OF TROPICAL RAINFOREST has been destroyed in the last 100 years.

414 FAST FACT...

AGROFORESTRY is a sustainable conservation technique, which protects soils from erosion.

415   DESERTIFICATION

DESERTIFICATION IS THE ACCUMULATED result of ill-adapted land use and the effects of a harsh climate. Four human activities represent the most immediate causes: over-cultivation exhausts the soil, overgrazing removes the vegetation cover that protects it from erosion, deforestation destroys the trees that bind the soil to the land and poorly drained irrigation systems turn croplands salty.

Moreover, the lack of education and knowledge, the movement of refugees in the case of war, the unfavorable trade conditions of developing countries and other socio-economic and political factors enhance the effects of desertification.

Desertification can lead to drought, which can have other geographical impacts and begins a vicious circle. Desertification when combined with drought conditions causes crops to failure, further soil erosion, famine, and hunger. People are then less able to work when their need is greatest and have to rely on food aid from the international community. What's really needed though is development aid, which involves educating the local community in farming practices.

416   FAST FACT...

THE SAHEL DESERT in Africa is expanding in size due to desertification.

417 RENEWABLE RESOURCES AND ENERGY

RENEWABLE ENERGIES ARE RELIANT on renewable resources including trees, fish, oxygen, fresh water, biomass, hydroelectric, solar, wind, wave, geothermal and tidal power, which, if managed in the right way, won't become exhausted. Renewable resources are vital to the development of a country so effective management is even more pressing for LEDCs.

Renewable sources of energy include hydro-electric power, wind, and solar power. As well as providing power, each of these has other benefits. Hydroelectric power for example, is most often associated with a dam that's built across a fairly narrow valley, causing a river that runs through it to back up on the upstream side and eventually form a reservoir, which promotes flood control, and can be used for recreation and as a wildlife sanctuary.

Alternative sources of energy are not without some criticism though. Some argue that tidal barrages could alter the look, behavior and ecology of estuaries and might cause pollution problems, and some people oppose wind farms on the grounds that they are ugly and noisy and a hazard to wild birds.

418 FAST FACT...

HYDROELECTRIC POWER supplies about a quarter of the world's electricity.

419 CARBON CREDITS

SOME MEDCs are trying to reduce their energy demand by introducing schemes such as Carbon Credits.

Aimed at reducing greenhouse gas emissions, the carbon credits system is based on the polluter pays principle according to how much pollution they generate. The idea is that people are encouraged to pollute less, as it will cost them less in carbon taxes.

A good example of this is the London Congestion Charge whereby drivers are charged for entering the congestion charge zone in central London during peak hours. The aim is to encourage the use of public transport, thereby reducing congestion, the time spent in queues, the pollution generated and the cost to the economy. The money generated is used to improve public transport, e.g., older London buses which generate more pollution have been removed from service. As a result of the London Congestion Charge, there has been a reduction in congestion, accidents and pollution levels, an increase in investment in public transport and retail sales inside the congestion zone have also increased.

420 FAST FACT...

📖 **ONE CARBON CREDIT** is equal to one metric tonne of carbon dioxide, or in some markets, carbon dioxide equivalent gases.

421 FAST FACT...

📖 **CARBON TRADING** is an application whereby greenhouse gas emissions are capped and then markets are used to allocate the emissions among the group of regulated sources.

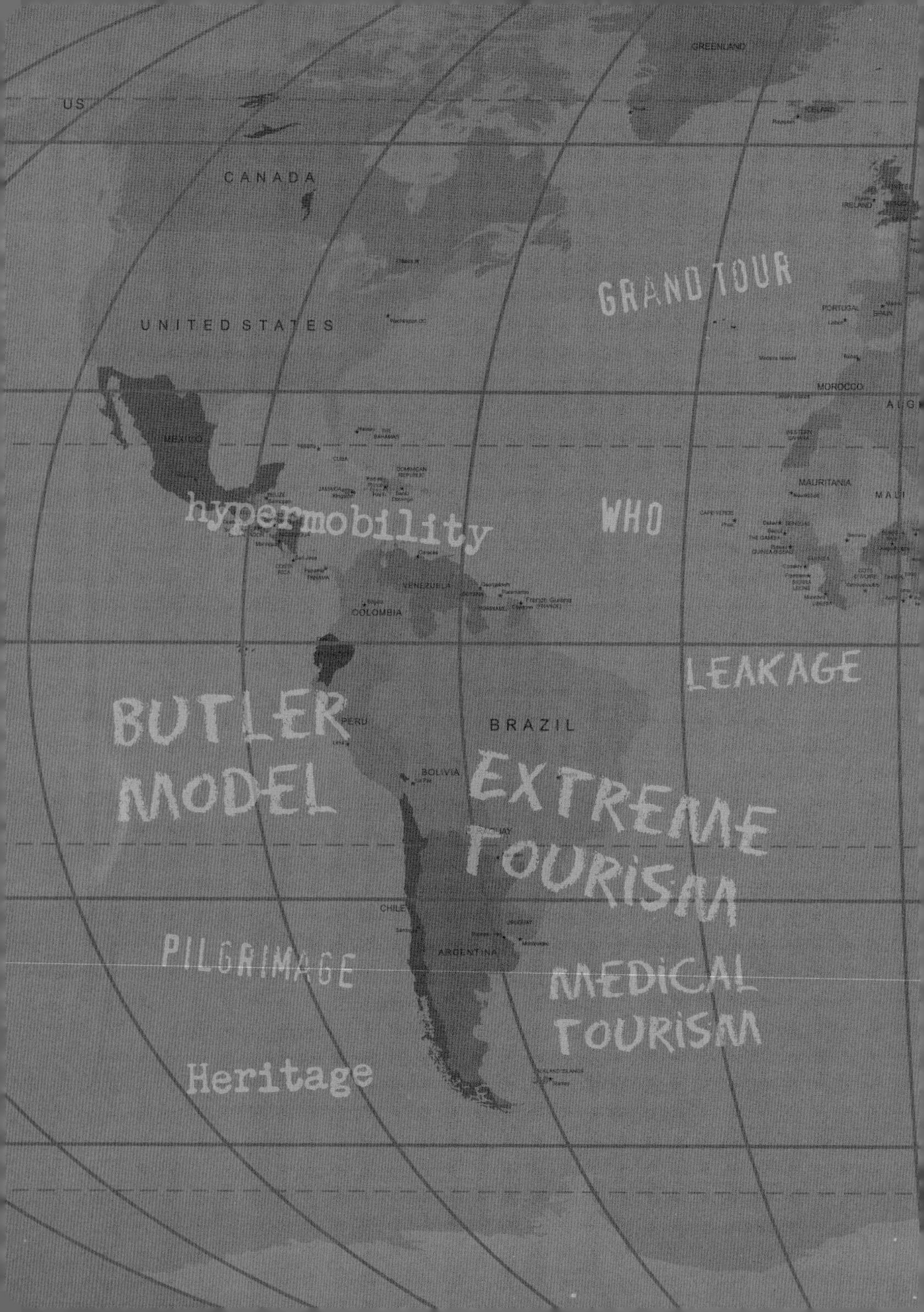

GREENLAND
US
ICELAND
CANADA
IRELAND
GRAND TOUR
PORTUGAL
SPAIN
UNITED STATES
MOROCCO
MEXICO
CUBA
MAURITANIA
MALI
hypermobility
WHO
VENEZUELA
COLOMBIA
LEAKAGE
BUTLER MODEL
PERU
BRAZIL
BOLIVIA
EXTREME TOURISM
CHILE
ARGENTINA
PILGRIMAGE
MEDICAL TOURISM
Heritage

501
SITES
CRUISES
Tourism
CULTURE
SAFARIS
PYRAMIDS
ENVIRONMENT
agri-tourism
RUSSIA
CHINA
JAPAN
IRAN
AFGHANISTAN
PAKISTAN
INDIA
BURMA
LAOS
THAILAND
PHILIPPINES
INDONESIA
AUSTRALI
LIBYA
EGYPT
SAUDI
ARABIA
SYRIA
JORDAN
TURKEY
UKRAINE
POLAND
CHAD
SUDAN
ETHIOPIA
CENTRAL AFRICAN REPUBLIC
KENYA
UGANDA
THE CONGO
DEMOCRATIC REPUBLIC OF THE CONGO
TANZANIA
MOZAMBIQUE
MADAGASCAR
NAMIBIA
BOTSWANA
SOUTH AFRICA

422 TOURISM

TOURISM IS TRAVELING FOR RECREATION, leisure or business purposes outside of your usual environment for short periods of time no longer than one year.

In recent decades, tourism has become a key leisure activity and a vibrant economy has grown around the industry. Some parts of the world are now dependent on tourism as a source of income and international tourism receipts grew to over one trillion dollars US in 2011.

The roots of tourism can be traced back to Roman times, but it was the mid 17th century when the 'Grand Tour' of Europe became popular amongst wealthy young men from England. Over time, the industrialization of the world and improved transportation systems allowed many more people to travel for pleasure.

Since then, many offshoots of tourism have developed in response to more niche demands. Ecotourism, dark tourism, and medical tourism are all recent innovations that allow travellers to take advantage of the particular environments, histories or facilities associated with a particular location.

423 FAST FACT...

📖 **ACCORDING TO THE WORLD**
Tourism Organisation, France is the country most visited by international travelers, attracting 79.5 million visitors in 2011.

424 TOURISM TRENDS

🎓 **ONE OF THE FIRST POPULAR** forms of tourism was exclusive to wealthy Europeans who would take a 'Grand Tour' as a prototype gap year, spending time with the aristocracy in neighboring countries, exploring new cuisines, and practicing languages.

Industrialization in the 1800s offered tourism to the masses with Thomas Cook popularizing the idea of package holidays by offering rail tickets and food for a combined reduced price.

Winter tourism arrived in the Alps in the 1860s, although it wasn't until the 1970s that Swiss resorts became more popular in winter than summer.

In the last few decades, tourism has increased due to there being more holidays from employers, cheaper travel, especially by air, and more money invested in the tourism infrastructure.

Tourism has continued to grow despite the financial crisis of the early 21st century, with improvements in IT and telecommunications making international communication common and travel cheaper.

425 FAST FACT...

📖 **HYPERMOBILITY** is a term coined in 1980 that describes the travel habits of people who cover much higher than average distances for work or leisure.

426 TOURISM GROWTH

🎓 **AS ONE OF THE WORLD'S** largest industries worth over one trillion dollars US every year, tourism is an important part of the economy for many countries.

It is also one of the fastest growing industries in the world with an overall increase of 37% worldwide in the last 10 years. Much of this growth has been in less economically developed countries, with Africa's industry doubling in a decade, South Asia's tourism increasing by 63% and the Middle East by 70%.

Tourism can boost economies and help fund infrastructure, but the cyclical nature of tourism can cause problems. Initially, a resort is 'discovered' by a few travelers and becomes popular. Next mass tourism arrives and overwhelms a resort as it becomes overcrowded and tourism destroys the attractive quality of the resort, leaving the area to go into decline.

In more economically developed countries, tourism continues to grow as travelers have greater working flexibility and cheaper transport, which has lead to special interest tourism.

427 FAST FACT...

📖 **THE FIRST SPACE TOURIST** was US citizen Dennis Tito who paid 20 million dollars US to orbit Earth for just under eight days.

428 FAST FACT...

📖 **TRAVEL AND TOURISM RANK** among the top 10 industries for employment in 48 states and Washington DC.

429 TOURISM FOR MEDCs

🎓 **TOURISM IS A VITAL PART OF THE ECONOMY** for lots of more economically developed countries as it helps to reduce unemployment and brings in large amounts of money for retail and service sectors.

Although tourism can help to regenerate an area, the increased pressure on the infrastructure can have drawbacks. In the UK, the regeneration of Bristol's harborside area was a successful way of reversing the near terminal decline brought on by the introduction of large cargo ships that could no longer dock there. The docklands have been redeveloped over the past three decades to create one of the largest redevelopment projects in Europe.

However, concerns have been raised that although it has created thousands of jobs and generated £300 million in internal investment, there are problems with congestion and the tourist attractions have driven up the cost of housing in the area, pricing local people out of the market.

430 FAST FACT...

📖 **BY 2015,** tourists will drive 60% of luxury purchases in Europe.

431 FAST FACT...

📖 **IN 2011,** 29.2 million people visited the United Kingdom, making it the world's 7th biggest tourist destination.

432 TOURISM IN LEDCs – ADVANTAGES

SOME LEDCs – the Caribbean, China, Kenya, Thailand, Malaysia, Mexico, and Egypt - have tourist industries, which are seen as a way of improving economies and quality of life for their people. Often, tourism is seen as a way of getting development started and governments invest (with the support of MEDCs) in the building of new airports, roads, and hotels.

As such, tourism in LEDCs is growing because tour operators are promoting holidays in LEDCs more, traditional holiday resorts like the Mediterranean have become crowded with noise, pollution and high costs and in LEDCs, overall holiday costs are lower.

There are many advantages of tourism for LEDCs including the creation of jobs, infrastructure, hospitals and schools as well as the growth of industries such as restaurants, taxi driving, gift making, travel guides, entertainment, and security.

433 FAST FACT...

LEDCs NEED a stable government if they're to succeed in tourism in the long term.

434 TOURISM IN LEDCs – DISADVANTAGES

 **THE HUGE SUMS** of money spent on tourism every year make up a significant portion of the economy for many LEDCs, but the effects can create as many problems as they solve.

The revenue and jobs created by the increasing number of tourists visiting LEDCs may bring positive financial input. However, if an overseas company is running a tourism business then it is likely that the management of that business is also foreign and may employ foreign staff to work there on a seasonal basis. If this is the case then a significant portion of the money generated may end up outside of the local economy.

Other disadvantages may include uncontrolled development, which can make for a unattractive landscape, the displacement of traditional activities (e.g., fishing) due to coastal development, importation of food, poorly paid employment for locals and use of valuable water resources.

435 FAST FACT...

📖 **THE ESCAPE OF MONEY** from tourism from a local area to outside parties is known as 'leakage'.

436 IMPORTANCE OF COASTAL MANAGEMENT

A HOLIDAY by the sea is the ideal getaway for many tourists stretching back through history. From the French Riviera to the Thai island paradise of Koh Phi Phi, coastal management has proven essential to attracting visitors and maintaining resorts.

A large part of coastal management involves sea defense, as one of the key attractions for tourists is the beach. Managing a coastline well requires an understanding of land use and balancing tourist demands against those of environmental groups, councils, and residents.

Management strategies can include hard engineering options such as building groynes – wooden barriers to stop beaches washing away – or soft engineering such as beach management where the sand is replaced. Both will preserve a beach to attract tourists but also have weaknesses such as initial expense or constant maintenance and coastal management is necessary to determine which is most suitable for a given area.

437 FAST FACT...

FOLLOWING THE 2004 Indian Ocean Tsunami, Koh Phi Phi island had 23,000 tons of debris removed to restore its coastline.

438 BUTLER MODEL

👨‍🎓 **AS WITH** the majority of products, tourist destinations have a lifecycle and in 1980, Butler created a model of tourist development that shows how a resort grows across seven separate stages.

Initial exploration begins with a small number of tourists in an unspoiled area with few facilities. Next, the involvement of local people provides facilities and a tourist season emerges. Advertisements appear and the area development is recognized as a tourist attraction. Growth starts to slow during consolidation but the area still attracts tourists, occasionally causing tension with locals.

439 FAST FACT...

📖 **THE FULL TITLE** of this model is the Tourism Area Life Cycle by Professor Richard Butler.

During stagnation, facilities may decline and the number of tourists may drop. At this point the model splits with two possible futures. Either the destination undergoes rejuvenation and visitor numbers may increase, or the area goes into decline causing job losses and a negative perception of the area.

It is worth remembering though that the Butler model is a generalization and will not apply to all resorts.

440 SPECIAL INTEREST TOURISM

🎓 **A GROWING** number of tourists are engaging in special interest tourism, which caters for people seeking an additional thrill, a more immersive cultural experience or a greater knowledge of their history as individuals and human beings.

441 FAST FACT...

📖 **DOOM TOURISM,** or 'Last Chance Tourism' is the practice of visiting environmentally or otherwise threatened before it is too late.

Adventure Travel can involve remote locations as well as extreme sports, and can include engaging with nature or a different culture by living in the same style as local people. Dark Tourism caters to the darker side of human nature and offers the opportunity to visit scenes of tragedy such as battlefields or the sites of natural disasters such as Pompeii. There are overlaps with Disaster Tourism – visiting a disaster area out of curiosity – and Ghetto Tourism, which attracts fans of music or art associated with a particular area such as rap music and South Central Los Angeles.

Motives for special interest tourism can vary greatly and include academics studying history or people coming to terms with personal issues.

442 FAST FACT...

📖 **ADRENALINE JUNKIES** can indulge in Extreme Tourism, which can include Chernobyl Tours and bungee jumping into an active volcano in Chile.

443 MEDICAL TOURISM

🎓 **TRADITIONALLY, MEDICAL PATIENTS** traveling overseas have moved from less developed nations to medical centers in highly economically developed countries for treatment.

But in recent decades, the rise of elective surgery and cheap travel has led to the emergence of medical tourism, which is now identified as a national industry in over 50 countries.

The roots of medical tourism go back thousands of years to the time of pilgrims traveling to visit the temples of healing gods. The Spa towns in the UK such as Bath were also popular for this as it was believed the mineral waters there promoted health.

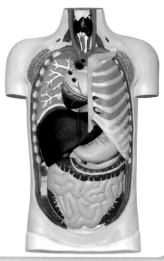

In modern times, the reasons for medical tourism are often based on convenience and price, for example a liver transplant in the US costs $300,000 but only around $91,000 in Taiwan. In more economically developed countries like the UK, medical tourists will travel to avoid long waiting lists.

444 FAST FACT...

📖 **IN 2007,** an estimated 750,000 US citizens went abroad for healthcare with the number expected to rise over the next decade.

445 FAST FACT...

📖 **FERTILITY TOURISM** is an offshoot of Medical Tourism where couples may travel abroad for fertility treatment due to legal restrictions in their own country.

446 RELIGIOUS TOURISM

🎓 EVERY YEAR, MILLIONS OF PEOPLE take part in Religious Tourism to make a pilgrimage, visit a country as missionaries or for fellowship. This practice, also known as faith tourism, has existed for thousands of years.

The biggest world event in Religious Tourism is the Muslim annual Hajj pilgrimage to Mecca in Saudi Arabia, which attracted over 2.9 million international pilgrims in 2011. In North America, Religious Tourism is valued at $10 billion US dollars a year.

The reasons for this type of tourism vary for individuals, but studies have found common reasons include people understanding and appreciating their religion through a tangible experience, to feel secure about their religious beliefs and to connect personally to a holy site.

As with other forms of tourism, Religious Tourism is on the increase as cheaper transport and greater personal freedom become available, according to the World Tourism Organisation, 300-330 million pilgrims visit key religious sites every year.

447 FAST FACT...

📖 **THE MOST FAMOUS** holy sites are the Church of the Nativity, the Western Wall, Brahma Temple at Pushkar, and the Kaaba.

448 WORLD HERITAGE SITES

THERE ARE SOME places in the world that as well as being tourist attractions are considered sites of outstanding cultural or natural interest. When these sites are at risk or in need of protection, the United Nations Educational, Scientific and Cultural Organisation (UNESCO) can designate them a World Heritage Site.

In 2012 there were 962 sites listed worldwide, 745 of them are cultural and include the Egyptian pyramids and Venice, 188 are natural and include national parks, 29 have mixed properties.

The first World Heritage Site was created in 1954 when the Aswan Dam was built in Egypt. Flooding the valley would have submerged several ancient treasures including temples, so these were moved and put back together further up the hill.

Each year the World Heritage Committee meets to decide whether new sites should be added to the World Heritage List.

449 FAST FACT...

📖 **OF ALL THE 157 COUNTRIES** with World Heritage Sites, the country with the most is Italy, home to 47, followed by Spain with 44, and China with 43.

450 FAST FACT...

📖 **WORLD HERITAGE SITES** are considered protected from acts of war under the Geneva Convention and international law.

451 CRUISES AND SAFARIS

🎓 **THROUGHOUT THE LAST 150 YEARS**, increasing numbers of tourists have focused on the journey itself as a tourist resort, leading to the increasing popularity of cruises and safaris.

Cruises first became available to the public in the latter half of the 19th century as ocean liners added luxury to the service and took longer routes further south during winter. Today, instead of ocean liners, specifically designed cruise ships offer greater capacity and comfort for cruise passengers, who numbered over 19 million in 2011 as part of a $30 billion dollar US industry.

A safari (originating from the Swahili word safari meaning long journey) on the other hand, is an overland journey often taking place in Africa. Modern safaris tend to focus on the ecological aspect of the journey rather than the opportunity for big game hunting, and the genre has expanded to include specialist safaris such as river safaris, balloon safaris, and safaris on horseback or elephant.

452 FAST FACT...

📖 **THE WORLD'S** largest cruise ships are the Oasis of the Seas and its sister ship Allure of the Seas, each with a gross tonnage of over 225,000 and capacity of 6,200 people.

453 IMPACT OF TOURISM ON CULTURE

LARGE AMOUNTS OF PEOPLE VISIT a new country on vacation for the opportunity to experience a different culture first hand and in doing so, expose that culture to their own way of life.

There are several positive effects of this interaction on cultures and on world society in general, as it educates people on a different way of life, helping to break down prejudices that are frequently the result of ignorance. Tourists visiting a new culture also serve to strengthen that culture as it is supported financially, as well as providing local employment for a population and reducing the need for them to migrate to find work.

Tourism can also damage a culture as traditions become marketable commodities for locals and often are adapted to suit the political or cultural views of the tourists. Possible side effects of tourism can include drugs and prostitution causing resentment amongst locals.

454 FAST FACT...

SEX TOURISM can lead to serious problems with child prostitution and disease.

455 FAST FACT...

IN SOME CASES, culture can be affected as young people look to foreigners as role models instead of local elders.

DEPENDING ON HOW IT IS MANAGED, tourism can have either a positive or negative effect on the environment as increasing numbers of people visit an area.

One of the main environmental benefits for a tourist resort is the ability to invest in protecting or restoring its natural resources and surroundings by using the additional financial resources. This is at the heart of sustainable tourism. Eco-tourism also attempts to reduce the impact of tourists by limiting numbers and using low-impact transport where possible.

When badly managed, tourism can have disastrous effects on the environment. The number of visitors to an environment without damage is known as the carrying capacity of an environment. This can be difficult to judge and requires careful monitoring by the authorities, as if left unchecked the tourists can end up destroying the reason for their visit.

457   FAST FACT...

IN PARTS OF SNOWDONIA National Park in Wales, large numbers of tourists have eroded footpaths leaving them six feet deep in places.

458   FAST FACT...

TOURISM MANAGEMENT has enabled the preservation of many historic buildings in Venice, Italy, enabling sustainable tourism.

459 SUSTAINABLE TOURISM

TOURISM IS A GROWING international industry that employs over 258 million people worldwide. In order to maintain such an important part of the economy, many tourism groups are looking to sustainable tourism as a way of ensuring future success.

Sustainable tourism is one of the core elements of eco-tourism, but while eco-tourism tends to focus on conservation and educating travelers on local ecosystems, sustainable tourism can be applied universally to all types of tourism.

At its heart, sustainable tourism minimizes environmental damage, maintains resources in terms of diversity, renewability and productivity and attempts to reduce the impact of tourism on local, regional, and global levels.

Within the travel industry, many companies are adapting to the demand for sustainable tourism. Many online travel agents allow eco-tourism and sustainable travel as search criteria, and hotels and resorts are adopting recycling and decreasing water and energy usage.

460 FAST FACT...

THERE IS CURRENTLY no international standard for either sustainable tourism or eco-tourism.

461 AGRITOURISM

🎓 **WITH THE RECENT** popularity in organic and grow-your-own foods, an increasing number of people are taking in interest in where their food comes from and how it gets to the customer. Agritourism is the adoption of this is as a business model by farms and food producers, enabling tourists to vacation on farms as well as buying products directly from farm shops and even trying their hand at harvesting and making regional produce.

Considered a growth industry in such parts of the world as Australia, Canada, the US and the Philippines, Agritourism has various definitions that range from staying at a bed and breakfast on a farm to picking fruits and vegetables by hand, riding horses or even staying on a cattle ranch, known as a dude ranch or guest ranch, and assisting in cattle drives.

462 FAST FACT...

📖 **WORLDWIDE OPPORTUNITIES** on Organic Farms (WWOOF), is a collective of national groups that place volunteers on organic farms.

463 ECO-TOURISM

FOLLOWING THE RISE of mass tourism in the 1970s and 80s, niche markets began to develop for special interest tourism such as Eco-tourism, also known as Green Tourism.

The term was popularized in the early 80s as a way of describing travel to fragile, naturally beautiful environments, and simultaneously protect them for the future. Frequently eco-tourists aim to travel to their destination using low environmental impact methods of transport such as cycling or walking, or by using mass public transport. This is in keeping with eco-tourism resorts, which aim to promote sustainable living and recycling as a way of life.

The main aim is to ensure tourists do not damage the environment in any way, however several criticisms of the concept have been raised as any extra visitors to a delicate ecosystem are likely to put pressure on the ecosystem. Additionally, without a central governing body there is a risk of exploitation.

464 FAST FACT...

IT'S ESTIMATED THAT the majority of eco-tourists – around five million – come from the US with the remainder from Western Europe, Canada, and Australia.

465 FAST FACT...

AN EXOTIC JOURNEY to a place 6,600 miles (10,000 kms) away uses around 185 gallons (700 liters) of fuel.

GREENLAND
US
CANADA
IRELAND
ECONOMIC
UNITED STATES
PORTUGAL
MOROCCO
MAURITANIA
MALI
social
POLITICAL
VENEZUELA
COLOMBIA
GREECE
HDI
PERU
BRAZIL
BOLIVIA
GDP
PARAGUAY
CHILE
ARGENTINA
STANDARD OF LIVING
RAMESSES III
KING DAVID

501

AID RUSSIA

TRADE

CHINA

fishing

IRAN

LIBYA EGYPT SAUDI ARABIA

SUSTAINABLE

INDIA

JAPAN

CHAD SUDAN

PHILIPPINES

ETHIOPIA SOMALIA

Development

INDONESIA

TANZANIA

INFLATION

UNEMPLOYMENT

MOZAMBIQUE

NAMIBIA BOTSWANA

MADAGASCAR

AUSTRALI

SOUTH AFRICA

ROSTOW'S
MODEL

WHO?

Green revolution

FORESTRY

UNDERSTANDING DEVELOPMENT

EVERY COUNTRY strives for development in order to be able to improve its economy and raise the quality of life of its people. Whilst development can begin with economic growth it also includes things like improved literacy rates, life expectancy, poverty rates, leisure time, environmental quality, and freedom or social justice.

There are four main areas of development:

Economic development – including greater income and wealth through industrial growth

Social development – better standards of living, access to education, health, housing and leisure environmental development – brings improvements and restoration of the natural environment

Political development – progress towards effective representative government development can have many different faces. For example, as Brazil develops its timber and mineral ores, parts of the Amazon Rainforest have been cleared for housing, agriculture and the Trans-Amazonian highway, thus raising the quality of life. In the Caribbean, the declining sugarcane and banana industries have been replaced by tourism, which has provided roads and power supplies as well as employment for the local population.

468 DEVELOPMENT INDICATORS

GEOGRAPHERS USE A SERIES of development indicators to measure a country's level of development. These include things like health assessment (does a population have access to medical care? If so, at what level and is it available to everyone?), industry evaluation (what type of industry dominates?) and education provision (does the population have access to education? If so, at what level and is it free?).

Gross National Product (GNP) – the total value of all goods and services produced by a country in one year, divided by the population total to give an average amount per head - is a key development indicator.

There are many criticisms surrounding the methods, including GNP, of measuring development though. GNP for example, doesn't include subsistence production for individual use, and within one country (from north to south or between cities and rural areas), there will be variations from the average national figure.

469 FAST FACT...

MORE ECONOMICALLY DEVELOPED COUNTRIES (MEDCs) are those with a high standard of living and a large GDP. Less economically developed countries (LEDCs) are those with a low standard of living and a much lower GDP.

470 FAST FACT...

THE HUMAN DEVELOPMENT INDEX was created by the UN in 1990 to provide international development comparisons.

471 HUMAN DEVELOPMENT INDICATORS

DEVELOPMENT OFTEN TAKES PLACE IN AN UNEVEN WAY. A country may have a very high GDP - derived, for example, from the exploitation of rich oil reserves - while segments of the population live in poverty and lack access to basic education, health and decent housing.

Human development indicators, measuring the non-economic aspects of a country's development, are therefore very important and include everything from life expectancy, infant mortality rates and poverty levels to literacy rates, level of access to technology, male/female equality and even government spending priorities.

Access to basic services, such as clean water and sanitation, necessary for a healthy life, as well as healthcare (taking into account statistics such as the number of doctors per patient, etc.) and risk of disease (related to conditions like AIDS, malaria and tuberculosis) are prime indicators of human development. Access to education, taking in the provision of primary, secondary and higher education, is also key.

472 FAST FACT...

IN THE UK, the infant mortality rate (the number of babies, per 1,000 live births, who die under the age of one year) is five, whereas in Kenya, it's 61.

473 FACTORS INFLUENCING DEVELOPMENT – PHYSICAL

THERE ARE MANY FACTORS which influence the rate at which a country may develop and they generally fall into either a physical or human category. Identifying and understanding these factors is important as it can change the focus for development.

Physical factors include:

Climate – the climate of a country can have a direct impact on the rate of development. The Sahel region in Africa for example, suffers from a lack of rainfall, which means that droughts are common and as a result crops suffer. There are also certain diseases such as malaria and yellow fever, which thrive in tropical climates because of the hot and humid conditions.

474 FAST FACT...

15 COUNTRIES IN AFRICA are landlocked, meaning it's difficult for them to trade, therefore affecting rates of development.

475 FAST FACT...

COUNTRIES, SUCH AS JAPAN, which are low in natural resources for example, have based their development on human factors such as education and skills.

Natural Hazards – not just droughts but floods and also tectonic activity can limit future growth and destroy buildings and agricultural areas. This also means a country may divert income to help recover from these events.

Geographical location – landlocked countries for example, experience difficulties in trade as goods have to be driven through other countries to get to the coast for shipping. It's also more difficult for new technology to reach a landlocked country, as fibre optic cables are laid under the ocean.

Natural resources – the presence of minerals, gas and oil can help improve a country's level of development but it's closely tied in with the ability to exploit the resource for the benefit of the country. Despite the advantage of MEDCs, there is interdependence between countries, which is vital to trade. For example, many LEDCs are dependent on MEDCs for manufactured goods or aid but MEDCs are dependent on LEDCs for primary products such as steel and iron.

FACTORS INFLUENCING DEVELOPMENT – HUMAN

FROM HISTORY to politics and economics to social factors, there are many ways in which human activity can affect development rates.

477 FAST FACT...

📖 **ONE IN SIX PEOPLE** don't have access to safe water, which means they're unable to work or care for their families because of illness, thus affecting development rates.

Historical events - Colonialism, for example, hindered a developing country's level of development. There was investment in colonies, but it was focused on things that would help the trade between the countries and borders of some colonial countries were set without attention to tribal and cultural differences, causing tensions and instability.

Politics – poor governance doesn't help a country to develop. Money that could be spent on development may be used to fund military weapons or an affluent lifestyle of an elite group of people, for example.

Economics – controlled by developed countries, world trade is uneven and unfair. LEDCs tend to sell primary produce but they have to compete against each other to win trade, which lowers the price for the farmer. Selling processed goods is more profitable but this is the domain of MEDCs.

478 FAST FACT...

📖 **FOREIGN INVESTMENT** can help a country to develop but it's very unevenly distributed. Africa for example, receives less than 5% foreign direct investment but has 15% of the world's population. Europe on the other hand, receives 45% of foreign direct investment, and has only 7% of the world's population.

Social factors – a less affluent country can't invest so heavily in a healthcare or education system. Without clean water for health or a basic education structure, development is stunted.

479 TRADE

TRADE IS THE MOVEMENT of goods and services between producers and consumers, and often between one country and another.

The pattern of world trade is a complex web of advantages and disadvantages. Usually, MEDCs export valuable manufactured goods such as electronics and cars and import cheaper primary products such as tea and coffee. In LEDCs the opposite is true, which means they have little purchasing power, making it difficult to pay off debts or escape poverty.

Increasing trade and reducing their balance of trade (the difference between the money earned from exports and the money spent on imports) deficit is essential for the development of LEDCs. Sometimes however, MEDCs impose tariffs such as taxes, and quotas, or limits, on imports.

480 FAST FACT...

GOODS BOUGHT into a country are called imports, and those sold to another country are called exports. Exports can include expertise and financial services.

Despite the advantage of MEDCs, there is interdependence between countries, which is vital to trade. For example, many LEDCs are dependent on MEDCs for manufactured goods or aid but MEDCs are dependent on LEDCs for primary products such as steel and iron.

481 FAST FACT...

A TRADE SURPLUS means that the value of exports is greater than imports. A trade deficit is when there are more imports than exports.

482 FAIR TRADE

👨‍🎓 **THE RESULT OF THE PATTERN OF WORLD TRADE** is that the workers in primary industries in LEDCs often lose out. They receive low wages and often have poor standards of living.

Fair trade means that the producer receives a guaranteed and fair price for their product regardless of the price on the world market. This means their quality of life should improve, as well as the long-term prospects for their children.

Fair trade products sometimes cost more in supermarkets in MEDCs, but many consumers consider this a small price to pay for the benefits they bring.

Fair trade sets minimum standards for the pay and conditions of workers. The Fair Trade Organisation promotes Global Citizenship by guaranteeing a fair, minimum price for products. In this way, they support producers in improving their living conditions.

Fair trade products are becoming more widespread and include tea, coffee, sugar, chocolate, and cotton.

483 FAST FACT...

📖 **ACCORDING TO THE FAIR TRADE ORGANISATION,** about 5 million people benefit from Fair Trade in 58 countries.

484 ECONOMIC DEVELOPMENT

THE ECONOMIC DEVELOPMENT OF A COUNTRY is widely seen as the key to an improved quality of life for its people, and is measured in a number of ways.

- Gross Domestic Product (GDP) - the total value of goods and services produced by a country in a year.

- Gross National Product (GNP) - the total economic output of a country, including earnings from foreign investments.

485 FAST FACT...

📖 **DEMOGRAPHICS** (population growth and structure) plays a key role in economic development.

- GNP per capita - a country's GNP divided by its population. (Per capita means per person.)

- Economic growth - the annual increase in GDP, GNP, GDP per capita or GNP per capita.

- Inequality of wealth - the gap in income between a country's richest and poorest people, which can be measured in many ways, (e.g., the proportion of a country's wealth owned by the richest 10% of the population, compared with the proportion owned by the remaining 90%).

- Inflation – by how much the prices of goods, services, and wages increase each year.

- Unemployment - the number of people who can't find work.

- Economic structure – the division of a country's economy between primary, secondary, and tertiary industries.

486 ROSTOW'S –MODEL

WALT ROSTOW bestowed his stages of economic growth model on the world in 1960. Based on the experiences of European countries and the USA, the model shows how these countries developed and was often used to explain the route that other countries would take. One of the more structuralist models, Rostow postulated that economic growth occurs in five basic stages, of varying length. These are:

1. Traditional society

2. Preconditions for take-off

3. Take-off

4. Drive to maturity

5. Age of High mass consumption

It is possible though, as indicated by The Indians who use their own rocket and satellite technology to track shoals of fish to help subsistence fishermen, to jump stages of economic development, i.e., by copying not reinventing the wheel. The Japanese also display this initiative who after the demolition of their industry at the end of the Second World War smartly spotted that they could do contemporary development their own way.

487 FAST FACT...

SOMETIMES CALLED the 'aeroplane model', each stage of Rostow's model has been illustrated with a different type of aeroplane. Pre-development: Kitty Hawk Flyer, Take-off: First World War Biplane, Acceleration: Boeing 747, Stabilization: Concorde.

488 GREEN REVOLUTION

🎓 **THE GREEN REVOLUTION** is the name given to a major form of development whereby the application of farming technologies gave a massive increase in food yields in low-income countries. Originally applied to agriculture in the Punjab region of India, the Green Revolution, also saw Indonesia for example, making efforts to increase the supply of rice and wheat using irrigation, new seeds, and fertilizers.

A new variety of rice, known as the IR-8, was created, that produced more grains of rice per plant and revolutionized rice farming. Newer high-yielding varieties (HYVs) have since reduced the growing period from 180 to 100 days.

489 FAST FACT...

📖 **ALTHOUGH THERE ARE** many seed types utilized in different places around the world in growing yields, the IR-8, remains synonymous with the Green Revolution.

The Green Revolution doubled total production in some countries, especially in China and provided higher standards of living as farmers were able to sell their surplus. It hasn't been without its problems though. Critics say that HYVs favor those farmers who can already afford fertilizers and irrigation and is to the disadvantage of the local-scale subsistence farmers with the gap between richer and poorer farmers, increasing.

490 SUSTAINABLE DEVELOPMENT

 **SUSTAINABLE DEVELOPMENT** meets the needs of the present generation whilst retaining our ability to meet the needs of future generations. Central to this is protection of the environment and natural resources.

491 FAST FACT...

📖 **GROUPS OF COUNTRIES** take international action towards sustainable development, which includes promotion, by MEDCs, of international trade to provide financial and technical assistance to LEDCs, and reduce poverty.

Sustainable development takes place through conservation, resource substitution, recycling, use of appropriate technology, pollution control, use of renewable energy sources, and lower consumption of energy through better insulation. And, examples of sustainable development include the use of brownfield land rather than fields for new housing, factories and business parks,

better wall and roof insulation as well as solar panels and recycled materials in new builds, the farming of forests so that more trees are planted than are being removed, the provision of cycle tracks in urban areas, improvements in electrically-driven public transport, organic farming and the use of alternative energy sources instead of fossil fuels.

492 FAST FACT...

📖 **SOME EXPERTS** believe that the world is moving towards unsustainable development as population increases alongside overconsumption of energy.

493 SUSTAINABLE FISHING

CURRENTLY, OVERFISHING – when fish are captured at a faster rate than they're able to reproduce – is creating an unstable marine ecosystem with unknown long-term consequences. Already, 90% of species at top position in the marine ecosystems food chain such as tuna, cod, sword fish and sharks have practically been eliminated or are in a situation of critical decline. And, if overfishing continues at the current rate, scientists estimate that many more fish will have become extinct by the year 2048.

Our oceans and the humble fish though is critical when it comes to feeding the world's growing population.

494 FAST FACT...

OCEAN OWNERSHIP, and therefore marine resource (fish!) ownership has led, and continues to lead, to conflict between countries where jurisdiction isn't clear.

495 FAST FACT...

GLOBAL WILD FISHERIES are believed to have peaked and begun a decline, with valuable habitats, such as estuaries and coral reefs, in critical condition.

Requiring no fresh water, producing little carbon dioxide, not using up any arable land, and providing healthy, lean protein at a more affordable cost than most meat, fish and therefore sustainable fishing have crept ever higher on political agendas.

Policies outlining quotas have been imposed as have regulations on net size and closed seasons as well as exclusion zones to conserve stock have all been put in to practice. Sadly though, illegal and unreported fishing accounts for an estimated 30% of global annual catches in recent years.

496 SUSTAINABLE FORESTRY

SUSTAINABLE FORESTRY allows some mature trees to be removed without decreasing biodiversity. The trees are cut, removed and replaced with new seedlings that eventually grow into mature trees. It's a carefully and skillfully managed system.

497 FAST FACT...

SUSTAINABLE FORESTRY is hard to supervise in some more remote forests but use of satellite technology and photography helps to check that any activities are following sustainability guidelines.

Sustainable forestry is essential to some countries, such as Brazil for example, which needs to exploit the Amazon's resources to develop. Left uncontrolled and unchecked exploitation of the Amazon could cause irreversible damage such as loss of biodiversity, soil erosion, flooding, and climate change.

Possible strategies for sustainable forestry include:

- Agro-forestry - growing trees and crops at the same time lets farmers take advantage of shelter from the canopy of trees. It also prevents soil erosion, and the crops benefit from the nutrients from the dead organic matter.

- Selective logging - trees are only felled when they reach a particular height. This allows young trees a guaranteed life span and the forest will regain full maturity after around 30 – 50 years.

498 FAST FACT...

A 'FOREST RESERVE' is an area that's protected from exploitation.

- Education - ensuring those involved in exploitation and management of the forest understand the consequences behind their actions.

499 AID

MEDCs **HAVE HIGH LEVELS** of economic development compared with LEDCs and therefore make allowance in their domestic budgets to provide aid to LEDCs. Many charities also exist to provide aid to LEDCs with the intention of helping them to continue development and improve quality of life of their people. There are many types of aid:-

- Emergency or short-term aid - needed after sudden disasters such as the 2000 Mozambique floods or the 2004 Asian tsunami.

- Conditional or tied aid - when one country donates money or resources to another but with conditions attached.

- Charitable aid - funded by donations from the public through organizations such as OXFAM.

- Long-term or development aid - involves providing local communities with education and skills for sustainable development.

- Multilateral aid - given through international organizations such as the World Bank rather than by one specific country.

There are many advantages associated with aid but also disadvantages for LEDCs such as dependency on MEDCs and that aid largely treats the problem but not the cause.

500 FAST FACT...

ORIGINALLY CALLED the United Nations Children's Emergency Fund, UNICEF was created to provide humanitarian assistance to children living in a world shattered by the Second World War. Today UNICEF remains dedicated to providing life-saving assistance to children affected by disasters.

501 FAST FACT...

THE CONTROVERSIAL Pergau Dam project in Malaysia, is an example of conditional or tied aid, where Britain used aid to secure trade deals with Malaysia.

INDEX

O

obesity 140, 142
oceans 23, 26, 35, 54, 78-101, 134, 136, 139, 173,
   243, 251
   ocean currents 89-90, 139, 149
   ocean floor 92, 95, 98
   oceanic plates 21, 94, 118, 121, 124
oil 44, 201-2, 243
   deep water oil 206
overconsumption 151, 163, 250
oxygen 19-20, 28, 30-1, 213
ozone 31

P

Pacific 46, 87, 91, 93
Pacific Ocean 51, 57, 78, 87-8, 93-4, 96, 163
petroleum 95, 202
photosynthesis 36, 38
planet 20, 23-5, 28, 30, 35, 144-6, 158, 162, 187
plants 16, 23, 27, 38-9, 42-3, 46, 86, 95, 168, 187,
   249
plates 116, 118, 121-2
poles 25, 139
pollution 141, 148, 182, 194, 204, 206, 214, 222
population 5, 27, 86, 142, 151, 153-8, 160-1, 163-4,
   168, 170-1, 174-5, 178-81, 201, 241-2, 247
pressure 26, 58, 123, 127, 130, 132, 135, 179, 236
   high pressure 43, 130, 133, 136
   low pressure 130, 135

Q

Quarries 204
quarrying 190, 204

R

radiation 31, 144
rain 39, 43, 57, 86, 89, 98, 129, 131-2, 135-6
rainforests 39, 200-1, 211
recycling 234, 236, 250
resources 65, 95, 162, 175, 200-2, 210, 234, 243,
   253
rivers 5, 23, 35, 38, 49-55, 57-68, 80, 88, 98, 105,
   110, 154, 173, 181-2, 213
Russia 44, 144, 153-4, 202

S

salt 85, 97-8, 129, 143
sand 45, 63, 75-9, 224
Savannah 37, 42
scientists 23, 25, 47, 51, 57, 104, 144, 149
sea 23, 26, 35, 45, 53, 56, 59, 65, 73-4, 77-9, 86, 96,
   98, 139, 231
   sea levels 15, 30, 80, 82, 99, 133, 139, 145, 173
Sedimentary rock 26
seismic waves 123
snow 32, 40, 44, 57, 105-6, 108, 114, 126, 129, 146
soil 23, 27, 38, 40-1, 57, 66, 86, 98, 154, 210-12
soil erosion 200, 212, 252
South America 42, 50, 88, 119, 178
South Poles 22, 25, 105
Spain 97, 175, 230

steel 198, 202, 243, 245
storms 75, 127, 135
stratosphere 19, 30-2, 133
Sun 20, 24-5, 30-1, 33, 38, 86, 98, 128

T

tectonic 18, 22, 120
temperatures 25, 28, 31-2, 44, 90, 107, 120, 125,
   128, 130, 133, 139, 144, 180
tornadoes 32, 137
tourism 5, 179, 211, 217-23, 229, 232-4, 240
   eco-tourism 233-4, 236
   sustainable tourism 233-4
transport 75, 81, 180, 191-3, 196, 198, 202-3, 236
travelers 220
trees 39-45, 64, 68, 187, 212, 250, 252
tropics 24, 89, 134
troposphere 18, 31-2, 133
tsunamis 47, 85, 92, 124
tundra 36, 44

U

ultraviolet rays 20, 31, 33
UNESCO 230
UNICEF 253
USA (United States) 89, 136-7, 144, 149, 165, 192,
   199, 208, 248

V

valleys 15, 21, 59, 60-3, 103, 110, 230
vegetation 37, 39, 41, 44, 64, 86, 139, 154, 212
velocity 59, 63, 80
Venezuela 63, 185
volcanoes 102, 116, 124-5
   volcanic eruption 92, 120, 124-5
   volcanic islands 46, 124

W

water 19-20, 23, 26-8, 35, 42, 50-3, 57-60, 62-7, 73,
   75-6, 86-8, 89-93, 98-9, 201-4, 208-10
   water cycle 85-6
waterfalls 49, 52-3, 55, 61, 63, 110
waves 71, 73-9, 92, 123, 213
weather 7, 31-2, 77, 98, 127-49, 206
Weber's model 195
Western Hemisphere 87
wind 28, 44-5, 73-4, 89, 93, 114, 121, 126, 134-7,
   139, 148, 201, 213
   wind speeds 134, 137
World Heritage Sites 230
World Tourism Organisation 218, 229

Y

Yangtze River 50, 61